Heterogenous Networks

LTE Coverage and Capacity Enhancement

Heterogenous Networks

LTE Coverage and Capacity Enhancement

Alex Fares

Heterogenous Networks

LTE Coverage and Capacity Enhancement

Published in USA.

Alex Fares

Heterogenous Networks

LTE Coverage and Capacity Enhancement

The Lord lives! Praise be to my Rock! Exalted be God my savior!

Psalms 18:46

To the memory of Phares Okumu Ndinya, and the great father he was.

Contents

Preface

Recent years have seen an exponential rise in wireless data traffic. Capacity demand rises far more quickly than spectral efficiency improvement, especially at hot spots/areas, due to heavily bandwidth-demanding applications like video and a new generation of devices (smart phones, netbooks, etc.). Additionally, more customers are working indoors as the service shifts from being voice-centric to data-centric, necessitating a higher connection budget and coverage extension to ensure a consistent user experience. Meeting demand in a cost-effective manner presents previously unheard-of issues for traditional networks designed for homogeneous traffic.

In previous generations, mobile networks were constructed using base stations, each of which had its precise position planned and the network structure adjusted for optimal service signal coverage. Cell splitting or the addition of carriers to the current network are two ways to increase the capacity in these kinds of homogenous networks, which exclusively employ conventional macro cells. Site acquisition has gotten more challenging as the number of denser networks has increased. Expanding this network design will not be a cost-effective way to satisfy the demand for more coverage and capacity. Therefore, "heterogeneous networks (i.e., HetNet's)" are crucial to the future of the mobile wireless industry because they require novel approaches to the network's overall cell-based design One feature of these HetNet's is their small base stations, which are simple to install inside an already-existing macro network and use modest power (between 100 milliwatts and 2 watts) to establish small coverage zones.

We provide an overview of small cells' ability to meet the high-capacity requirements of in-building users in mobile networks in this book. In the process, we go over pertinent theoretical underpinnings, important enabling technologies, and three main approaches to enhancing network capacity: network densification, spectrum accessibility, and spectral efficiency (SE) enhancement. Many Mobile Network Operator (MNO) Small Cell Base Station (SBS) topologies are assessed for spectrum accessibility. Cognitive radio approaches are assessed for Dynamic Spectrum Sharing (DSS) among several MNOs in a country to increase SE. For a particular Co-Channel Interference (CCI) limitation, the spectrum reuse is assessed for the network densification at both the intra- and inter-building levels.

Demanding work has been put together to produce this volume, there is little in this edition, which does not belong and cannot be used. A few things of consequence may have been omitted. But it is much easier to add than to cut, additions, which do turn out to be valuable, will be made in later editions. Technical difficulties of this book vary with the readers' background. Although the book is designed to cover Heterogenous Capacity Enhancement at specialist level, the non-specialists will find this book extremely helpful as a resourceful text. They will find this book to be in its own class in terms of consistency, illustration and practicality.

The unified comprehensive approach used in this book will appeal to Cellular Network Engineers, Cellular Network Operations Engineers, Regulatory Engineers, Network Integrators, Network Optimization, Decision Makers, Heterogeneous Network Designers and system suppliers who need to operate troubleshoot or operate Heterogeneous Networks at radio access level. This book will equally be an invaluable resource for Telecommunications students studying Network planning, Operations and Integration.

Abbreviations

1G	First Generation Wireless Networks
2G	Second Generation Wireless Networks
2D	2-Dimensional
3D	3-Dimensional
3G	Third Generation Wireless Networks
3GGP	Third Generation Partnership Project
4G	Fourth-Generation Wireless Networks
5G	Fifth-Generation Wireless Networks
ABS	Almost Blank Subframe
APP	ABS Pattern Period
BS	Base Station
CoMP	Coordinated Multipoint Transmission
CP	Cyclic Prefix
CCI	Co-Channel Interference
CDMA	Code Division Multiple Access
CINR	Carrier-to-Interference-and-Noise-Ratio
CR	Cognitive Radio
CSA	Co-channel Shared Access
CSI	Connectivity Service Network
cSBS	Co-channel SBS
DSS	Dynamic Spectrum Sharing
Femto ABS	Femtocells Advanced Base Station
EE	Energy Efficiency
eICIC	Enhanced Intercell Interference Coordination
hRF	Horizontal Reuse Factor
ICI	Inter-Carrier Interference
IMT	Advanced International Mobile
IP	Internet Protocol
IPSec	Internet Protocol Security
ITU	International Telecommunications Union

LAA	Licensed Assisted Access
LOS	Line-of-Sight
LTE	Advanced Long-Term Evolution Advanced
LSA	Licensed Shared Access
MR-BS	Multihop Relay Base Station
MBS	Macrocell Base Station
MS	Mobile Station
mmWave	Millimeter-Wave
MNO	Mobile Network Operator
PHY	Physical Layer
p-MNO	Primary MNO
PU	Primary UE
QCI	QoS class Identifier
QoS	Quality of Service
RAT	Radio Access Technologies
RoE	Region of Exclusion
RS	Relay Station
SBS	Small Cell Base Station
SC-FDMA	Single Carrier Frequency Division Multiple Access
SE	Spectral Efficiency
SIB	System Information Block
SINR	Signal-to-Interference-plus-Noise Ratio
SLSA	Static Licensed Spectrum Allocation
s-MNO	Secondary MNO
sSBS	Serving SBS
sSU	Serving Small Cell UE
SU	Secondary UE
TTI	Transmission Time Interval
UE	User Equipment
UpPTS	Uplink Pilot Time Slot
vRF	Vertical Reuse Factor

1

Motivation For Cellular Heterogeneous Networks

We live in a world that rapidly changes in terms of technology, social architecture, knowledge, spirituality, and innovation. These changes have taken a centerstage and influence how we create wealth, how we socialize, how we bring up and discipline our children, how we play and how we conjecture our future. Telecommunications users today desire to interact with one another using any medium, such as email, instant messaging, audio, and video. Users utilize their smart mobile devices to create and consume material created by themselves or by commercial media, and they wish to use social networking to share their personal life experiences, thoughts, and news with others. Meanwhile, mobile internet is quickly moving from people to machines and becoming embedded with internet [1].

The rapid growth of the small cell market is being fueled by the development of several key technologies. Moore's Law continues to offer the semiconductor industry more transistors and power budget, enabling them to produce enhanced signal processing capabilities at reduced costs and optimized power consumption. With applications moving from voice to data and video to augmented reality, user interfaces changing from keypad to touch to gesture, application developers are constantly innovating and optimizing the advantages of signal processing technologies. To meet the enormous demands for capacity and bandwidth, our communication infrastructure must also adapt as our society enters the era of "Big Data." This is imperative. Therefore, to accommodate a wide range of connectivity and provide an unparalleled user experience, the transition from homogeneous to heterogeneous network design is imperative.

In both scholarly study groups and the wireless industry, heterogeneous networks have drawn a lot of interest as one of the key foundations and future directions of mobile communication technology. The goal of this chapter is to present a business and technological overview of heterogeneous networks, the current stage of technological development, the primary trade-offs and obstacles, and the directions for future research and development. We anticipate that as information technology and telecommunications continue to merge and mobile devices become more sophisticated, heterogeneous network technology will also continue to evolve.

1.1 Heterogeneous Networks Motivation and Definition

In today's wireless communications market, one of the most used but least clearly defined terminology is heterogeneous networking. The overlay of small cell networks (such as micro, pico, and femtocells) and macro base station networks are sometimes referred to as heterogeneous networks. Others view a primary use case that combines a Wi-Fi network with a cellular network. Some people also think that the vision of heterogeneous networks should include novel network topologies and connectivity, like traffic aggregation for machine-type devices, peer-to-peer, device-to-device, personal hotspot, relay, and near field communication (NFC). It is possible that cognitive radios will be incorporated into heterogeneous networks once the network infrastructure integrates flexible sharing and dynamic spectrum access.

Notwithstanding these differing interpretations and definitions, research and implementation of heterogeneous networks have advanced significantly in recent years, especially in the field of data offloading via small cells. Mixed deployments with macro, pico, femto, and relay nodes are referred to as heterogeneous deployments. Heterogeneous networks typically have several tiers (or layers) of networks with varying cell sizes and footprints, as well as diverse radio access technologies [2]. Multi-tier heterogeneous networks are exemplified by an LTE macro base station network that sits on top of an LTE pico base station network. Since the same LTE air interface technology is utilized in various network layers and tiers, 3GPP, the standards organization that developed LTE, has produced design provisions and solutions to make it easier for such a heterogeneous network to interact and integrate. One such solution is a comprehensive performance evaluation methodology that simulates a range of deployment scenarios.

However, an excellent illustration of a multi-tier and multi-air-interface network is a heterogeneous network that consists of a Wi-Fi network and a macrocellular LTE network. The heterogeneous network requires cooperation between standards development groups because the air interfaces were created by several entities (in this case, IEEE802.11/Wi-Fi Alliance for Wi-Fi__33 and 3GPP for LTE). One such example is the Wi-Fi Alliance's Hotspot 2.0 specification. Future heterogeneous networks might have extraordinarily large base stations in addition to small cells. One example is Cloud-RAN [3]. A cloud-RAN base station moves all or most of the baseband signals from tens to hundreds of traditional stations to a centralized server platform for enormous signal processing using high-speed optical fiber links.

Since air conditioning may no longer be necessary at each cell site, this architecture may drastically lower the network's energy consumption. Additionally, a super base station's scale allows it to dynamically distribute its signal processing resources to adjust to the fluctuating traffic loads within its geographic coverage during the day phenomenon known as the "tidal effect" "These further lowers energy consumption and hardware requirements. These super base stations might make it easier for various base station kinds, such as macro and small cells, to be coupled more tightly. To boost network efficiency, they might also provide a foundation for the economic deployment of sophisticated air interface and network features, which will go into greater depth of innovation. There is a common set of challenges and techniques to integrate various wireless technologies together to form a high-performance heterogeneous network, even though the size of base stations in each tier or layer of a hybrid network can vary greatly, ranging from femto stations to picocell, microcell, macrocell, and cloud-RAN. Every tier may use different radio access technologies.

It should be noted that due to the broad technical scope and highly complex economic tradeoffs, the designs of a heterogeneous network may have different flavors, concentrations and motivation, depending on the existing installed network equipment, the choices of network transport, the availability of the required multi-mode devices, and the main set of applications expected to be supported by a particular heterogeneous network. As the requirements from users and mobile network operators continue to evolve, the definitions and the technical concentrations of heterogeneous networks are expected to also evolve with time.

1.1.1 Enablers for Heterogeneous Networks

The fast growth of heterogeneous networks can be linked to essential technological and economic factors. In coming years telecommunications are anticipated to be significantly impacted by the results of this technical advancement as enhanced by Heterogeneous Networks. Since, Heterogeneous networks are driven by the necessity to improve network performance, efficiency, and adaptability in varied environments. We expect to see improved coverage, capacity, and user experience by merging different cell types (macro, micro, pico, etc.) to improve mobile network performance. This includes improved service quality, decreased congestion, and quicker upload and download speeds. By maximising use of resources and facilitating more effective network operation, HetNet seek to meet the growing demand for data and connection. However, motivation factors attributing to rapid deployment of Heterogeneous networks are as outlined below.

1.1.1.1 Challenges Serving Indoors with outdoor Macrocell Base Station (MBS)

To support rich multimedia services on mobile phones, a significant amount of data is created by indoor users at high data rates on typical cellular mobile networks. This is especially true in metropolitan high-rise buildings, many of which house hundreds of units. It is challenging to serve this large amount of indoor data at a high rate with an outdoor Macrocell Base Station (MBS) due to a building's high external wall penetration loss, a limited system bandwidth below 3 GHz, and a maximum transmission power that limits excessive interference.

Therefore, addressing the huge data rates and massive capacity demands indoors is now unavoidable. In this sense, installing Small Cell Base Stations (BSs) within buildings is thought to be an efficient way to serve a significant volume of indoor data at a high data rate due to the limited coverage and low transmission power. Shannon's capacity formula identifies three primary techniques to increase network capacity: network densification, spectral efficiency augmentation, and spectrum accessibility.

1.1.1.2 High-frequency-High-Capacity Millimetre-Wave (mm Wave) spectrum already Utilised.

The high-frequency Millimeter-Wave (mmWave) spectrum bands have already been availed to address the high-capacity demand of Fifth Generation (5G) and advanced mobile systems, especially indoors within multistorey buildings. This is due to the near-complete occupancy of spectrum bands below 3 GHz. Therefore, the short-range and abundance of mmWave spectrum are a promise for addressing the large-scale deployments of small cells to provide high data rates at a short distance, especially in urban indoor coverage. To further expand the spectrum, more spectra can be made available, allowing each tiny cell to function in many spectra.

1.1.1.3 Inter-cell Interference Coordination (eICIC gets simpler

In addition, each nation's Mobile Network Operator (MNO) is often given a specific licensed spectrum to manage user traffic. In the past, these static radio spectrum allocations were adequate to meet consumer demands. The dynamic sharing of radio spectrum that has already been statically assigned to MNOs employing indoor small cells has recently been introduced to be more effective for reusing the same dedicated spectrum for an MNO. Dynamic Spectrum Sharing (DSS) allows a system (primary) to share its spectrum statically or opportunistically with another system (secondary), provided that the primary system is not impacted by the sharing. By making use of their layouts, indoor smaller cells become highly significant in implementing various DSS techniques. Enhanced Intercell Interference Coordination (eICIC) based on Almost Blank Subframe (ABS) can be utilised in small cells to allow time orthogonality while servicing traffic of the corresponding MNO able to eliminate Co-Channel Interference (CCI) when sharing the licensed spectrum of two MNOs.

1.1.1.4 Spectral Efficiency Improvement by Use of Cognitive Radio (CR)

By using Cognitive Radio (CR), MNOs in a nation that has difficulties in facilitating the effective use of its available licensed spectrum can improve spectral efficiency. This is because the need for the necessary quantity of spectrum for various MNOs varies dramatically over time and space due to the abrupt changes in the user traffic demand of various MNOs in each nation. This results in the underutilization or unused time or space of a significant amount of the spectrum allotted to each MNO in a nation. To solve this issue of underutilization of spectrum, Cognitive Radio (CR) has emerged as an enabling technology in recent years. Spectrum access plays a key role in CR by preventing collisions between the primary and secondary User Equipment (UE). This enables the sharing of licensed spectrum between MNOs, increasing their effective spectrum bandwidth and enhancing their spectral efficiency to support high capacity. Through resource allocation optimization, adaptive spectrum management, and intelligent spectrum management, Cognitive Radio (CR) can improve spectral efficiency in small cell networks. When main users (PUs) are not using the spectrum, CR enables secondary users (SUs) to access it. This dynamic spectrum access (DSA) improves radio frequency spectrum utilization.

1.1.1.5 Spectral Efficiency Improvement by Use of the intelligent reflecting surface (IRS)

An innovative radio technology for improving wireless communication systems the intelligent reflecting surface (IRS). IRS, a planar meta surface made up of several reflecting elements, has received a lot of attention to increase the spectral efficiency of future heterogeneous wireless networks. An IRS is a meta-surface that uses many tiny, programmable, passive, inexpensive reflecting parts to impart a controlled individual phase shift to the impinging electromagnetic wave. It is also referred to as a large intelligent surface, software-controlled meta-surface, or reconfigurable reflecting surface. To optimise signal intensity and reduce interference, the propagation characteristics between the transmitter and receiver are modified [3]. While IRS creates a virtual link between the transmitter and receiver, causing the receiver to achieve the transmitted signal quality and thereby improving the overall SE (Spectral Efficiency), with line-of-sight (LOS), IRS allows the creation of a replica copy of the original LOS signal to increase the system's diversity.

1.1.1.5 Spectral Efficiency Improvement by Use of Non-Orthogonal Multiple Access (NOMA) techniques

Non-Orthogonal Multiple Access (NOMA) techniques are also a potential option to meet the high spectral efficiency requirements of networks beyond 5G, in contrast to Orthogonal Multiple Access (OMA) schemes, which are intrinsically inefficient because they limit the number of users to be served simultaneously. More particularly, NOMA has lately been regarded as a crucial technology for Beyond 5G networks, which significantly increase spectrum efficiency by allowing numerous users to utilize time and frequency resources simultaneously [4]. Code Domain NOMA (CDNOMA), which allocates users distinct spreading codes, and Power Domain NOMA (PD-NOMA), which uses different user power levels to share time, frequency, or code slot simultaneously, are the two basic NOMA methods [4].

1.1.1.6 Explosive Growth of Data Capacity Demands

User demand and complexity for data capacity on mobile internet has skyrocketed in recent years. The extensive use of progressively sophisticated mobile devices is foremost contributor to this. According to market research, the size of the device's screen, how user-friendly its operating system is, and how responsive the wireless network is connected to, directly influence the volume of data flow. For instance, a tablet uses five times the system capacity of a smartphone, and a 3G smartphone typically uses around thirty times that of a 2G voice phone. The need for data capacity will keep growing as mobile devices' screens, image quality, and battery life continue to improve, and as network infrastructures' peak data rates and latency continue to improve.

Along with this expected increase in capacity, the network is also subject to a considerable additional load from content created by users, social networking, and the demand from upgraded mobile devices, communication infrastructure signaling and monitoring. Since mobile devices provide extensive coverage with their always-on and always-connected accessibility, they are the perfect platform for social networking applications. Short yet periodic data transmissions are typically produced by social networking and related applications. To support many users, a network may need to regularly establish and disconnect radio links to preserve valuable radio resources. This frequently leads to a flood of control messages being sent over the control plane. Nevertheless, as the prevalence of viewing YouTube videos, TikTok on mobile devices grows, so does the capacity customers demand on the data plane, which is frequently growing asymmetrically between the uplink and the downlink. Lastly, cloud-based services may further accelerate demand as information is transferred between mobile devices to the cloud for network storage and cloud-based computing, cantered upon the way the client device and cloud allocate the processing of signals load.

Network congestion is also beginning to spread from the last mile access network, which is predictable traffic processes bottleneck, to the core network, and eventually to the backbone network and its connections. Future wireless network series of generations will become even extra complex due to machine-type communications. Future networks are poised to grow and complexity as mobile internet transitions to embedded internet to manage unprecedented, linked devices with wildly disparate usage patterns, traffic patterns, and security needs.

These machine-type devices must be able to manage a wide range of capacity needs, from high-speed real-time surveillance footage to extremely low traffic volume monthly meter readings. The complexity of capacity planning is further increased by the need to safely manage billions of these linked devices across a wide range of networks and operational environments. Future wireless networks will need to boost their capacity by orders of magnitude due to the combined needs of social networking, user-generated content, organic traffic growth, and machine-type connected devices. The rise of heterogeneous data traffic also necessitates a paradigm change in the design and provisioning of network architectures.

1.2 From Spectral Efficiency to Network Efficiency Challenges

The rapid expansion of data traffic can be addressed in several ways by the wireless sector. We are currently at the theoretical limit of radio channel capacity, also referred to as the Shannon limit, following decades of advancements in air interface technology. The change from concentrating exclusively on spectral efficiency to taking network efficiency into account recognizes that a successful network depends on more than just maximizing data transmission per unit of bandwidth (spectral efficiency). Network efficiency includes more comprehensive elements including total network throughput, resource utilization, and user experience. For an even higher spectral efficiency, the mobile communications industry requires several orders of magnitude in relation to more system capacity than what air interface spectral efficiency improvement can provide.

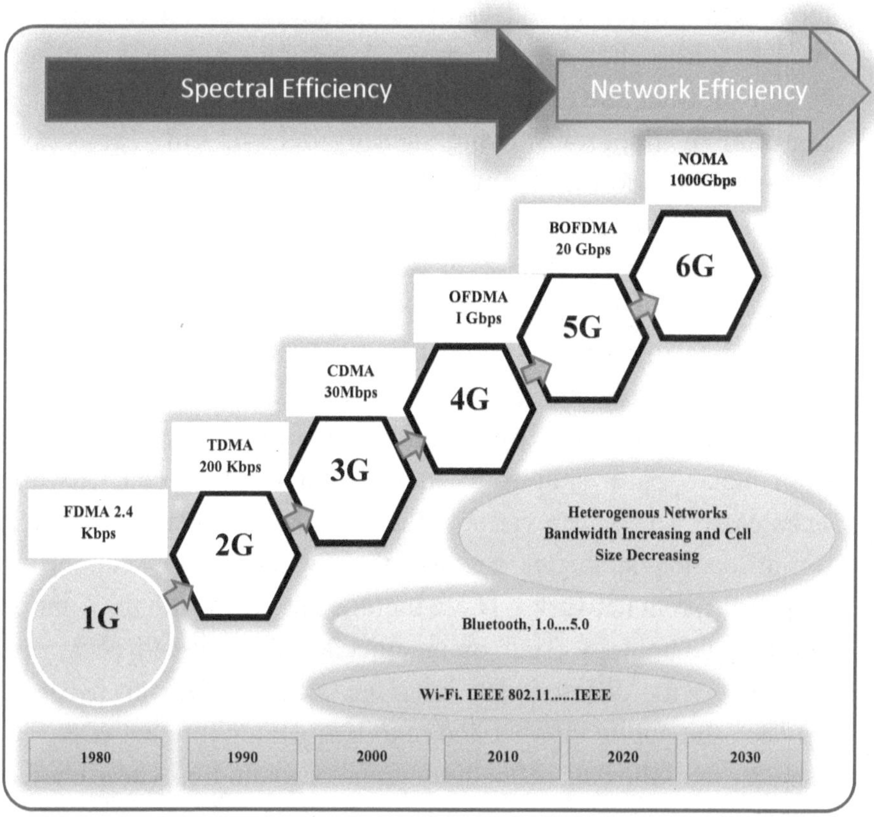

Figure 1.1 Wireless technology evolution

Therefore, rather than depending only on spectral efficiency improvement at the radio link level, future capacity growth must come from a combination of technological solutions, including optimizing the overall network efficiency.

1.2.1 Data offloading Initiatives

Mobile network providers are considering restricting each subscriber's monthly data use over wireless wide area networks (WWAN) as a short-term measure. When required, they will also throttle the data rate of users who use the network excessively. But constraining usage or slowing capacity demand only serves as a short-term solution to the current network overloading issues. To support and facilitate future sustainable growth in data traffic, give all users access to mobile broadband, and improve everyone's quality of life, we require more initiative-taking solutions.

The data offloading approach is one such option that mobile network operators are considering. This creates a fundamental heterogeneous network by enabling users to transfer their traffic from macro base stations to the alternative small-cell networks seamlessly. There are a variety of technical options and preferences for offloading strategies because the capacity bottleneck differs within different networks. These include networks with the same air interface technology (such as macrocell and small-cell networks), networks with different air interface technologies, or networks between the public internet and mobile operator core network. The subject of mobile data offloading has no one solution. These choices complement one another, and they will all keep evolving to satisfy the ever-rising capacity requirements.

The addition of more spectrum is another understandable solution to the increasing demand for data capacity. The wireless industry and regulators are collaborating to explore the possibility of adding additional licensed and unlicensed frequency bands for mobile internet applications; however, the progress in new frequency allocation has been slow due to the restricted availability of spectrum and an overwhelming need for globally harmonized allocation to maximize the economy of scale. There is more pressure to re-farm the current frequency bands and to quickly deploy small cells for high spatial frequency reuse because many highly populated markets are already on the edge of running out of spectrum. Furthermore, for mobile internet applications, the wireless sector has begun to examine high frequency bands like millimeter wave. The technology, design, and operation of these networks are anticipated to differ significantly from those of traditional cellular networks due to the distinct radio propagation characteristics of these bands compared to the traditional lower frequency bands (typically below 3 GHz) utilized for high mobility cellular networks.

Consequently, the primary means of attaining increased system capacity are *heterogeneous networks*, which are composed of layers of networks operating at several frequency bands. The wireless industry and academic circles are exploring novel approaches for enhancing flexible spectrum utilization, such as spectrum sharing, dynamic spectrum access, and cognitive radio with opportunistic network access, in addition to securing additional spectrum allocation and creating innovative technologies for higher frequency bands.

TV white space is defined as unused portions of the radio frequency spectrum in the television broadcast bands. This "white space" can be utilised to provide additional wireless services, such as broadband internet, without interfering with existing TV transmissions. These unoccupied spaces occur due to the geographical distance necessary between TV stations transmitting on the same channel. The use of the TV white space spectrum for wireless communication in the US market is one instance of this. When the licensed service is not using the broadcast television spectrum, this new form of spectrum access necessitates the deployment of extra network entities, such as databases that manage the alternative radio transmitters. Due to the dynamic nature of spectrum availability, network coverage, and service availability differ from those of traditional wireless mobile networks. As a result, the industry continues to explore appropriate network architecture and business models to attain a sustainable return on investment. These new kinds of networks are anticipated to join the global heterogeneous networks from the perspective of telecommunications infrastructure.

1.2.2 Challenges in Service Revenue and Capacity Investment

Data services have replaced circuit-switched voice and short message service (SMS) as the primary drivers of mobile service revenue growth in recent years. For three primary reasons, this change raises the pressure on mobile network operators' profitability.

1. In comparison to SMS and conventional voice services, mobile data generates less income per bit.
2. Over-The-Top mobile applications are fiercely competing with the extremely lucrative operator walled-garden mobile applications.
3. MNOs must make a significant capital investment in additional network capacity to match the demand due to the explosive growth in mobile data traffic.

Because mobile network operators play a key role in funding, running, and maintaining the world's mobile internet infrastructure, it is imperative that the wireless sector and academic research communities create new networking innovations that enable operators to stay competitive as well as profitable so they can keep funding capacity and develop novel offerings. One of the key technologies that not only increases system capacity by tens to thousands of times but also makes it possible for new service generations to replace the declining revenue from traditional voice-centric telecom services is heterogeneous networking.

The future focus of wireless technology must shift from further increasing the spectral efficiency of the radio link to improving the overall network efficiency through heterogeneous network architecture and related signal processing technologies. This is because the demand for data capacity is increasing rapidly and the improvement in spectral efficiency in homogeneous networks is slowing down due to the approaching Shannon limit.

The Shannon limit, also known as channel capacity, is the theoretical maximum rate at which information may be successfully transferred over a communication channel with a given level of noise. It is a key idea in information theory, first introduced by Claude Shannon in his 1948 paper "A Mathematical Theory of Communication". This limit is determined by the channel's bandwidth and signal-to-noise ratio (SNR). the Shannon limit is often expressed using the following equation:

$$\mathbf{C = B * \log_2(1 + S/N)}$$

Where:

- **C**: is the channel capacity (bits per second)
- **B**: is the bandwidth (Hz)
- **S**: is the received signal power
- **N**: is the total channel noise power within the bandwidth B

This formula makes it clear that raising the SNR or bandwidth will raise the channel capacity. But according to the Shannon limit, there is a maximum data rate that can be achieved for a particular channel, and no amount of ingenious engineering can raise this limit. This means that the bigger the channel bandwidth and the higher the signal-to-noise (SNR) ratio, the higher the achievable data rate. This formula establishes the theoretical maximum data rate, which is obviously not fully realised in practice. There are no restrictions placed on how low the achievable error rate can be.

Depending on the coding technique employed, that is. This makes it possible to determine the minimum SNR needed for data transmission. This happens as the available bandwidth approaches infinity and is referred to as the Shannon limit. Increasing traffic density requires a higher system capacity, which heterogeneous networks may provide. Enhancing service consistency and fairness across the network coverage areas, especially at the cell edge, utilizing heterogeneous network architectures should be a focus. Additionally, we hope to leverage heterogeneous networks as a foundation for future technical advancements, such as the incorporation of new network architectures, connectivity models and interfaces, linked devices and applications.

1.3 Heterogeneous Networks Economics

Heterogeneous networks present unique economic advantages as well as challenges due to their varied components and functionalities. These networks, which are common in domains like social structures and mobile communication, have the potential to boost productivity, creativity, and resource allocation.

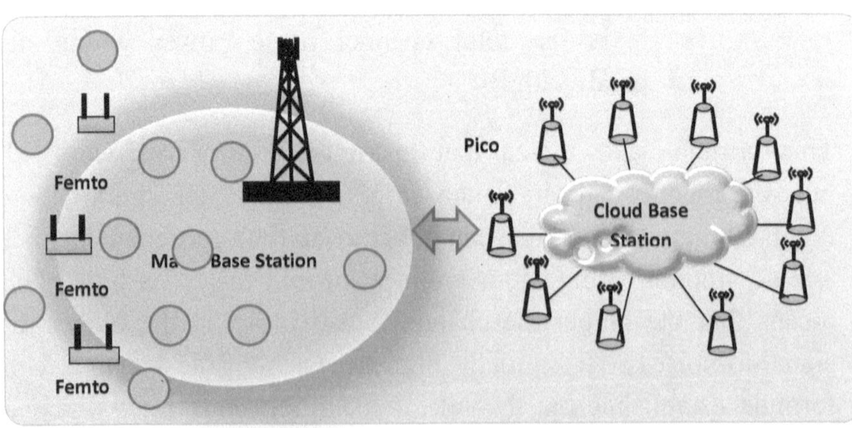

Figure 1.2 Femto, Macro and Pico Base stations

They do, however, also add complexity to pricing, management, and resource allocation. The economics of heterogeneous networks consider a variety of aspects, including the overall cost of ownership for mobile network operators, end user aquation cost, maintenance and operations costs, administrative costs, and Capital Expenditures. Key features of the economics of heterogeneous networks are outlined below, however there are several economic related challenges with regards to Heterogeneous Networks:

a. **Complexity and Management:** It can be expensive and difficult to integrate and manage various technologies and components; this calls for advanced management systems and specialised knowledge. Heterogeneous complexity systems comprise many components, each with specific features and functions. These components may consist of people in general, procedures, technology, or even several components of a product. The extensive web of interactions between each of the elements and the constituent components themselves are the sources of complexity. The elements and their interactions in heterogeneous systems are frequently dynamic introducing level of challenging complexities.

b. **Pricing and Cost Allocation:** Determining fair pricing and cost allocation mechanisms for services delivered across different technologies can be challenging, particularly when users access multiple technologies simultaneously. When coupled, heterogeneous pricing where many different customers or market segments pay varying prices and varying cost allocation the process of assigning expenses to specific services or operations can pose serious problems. These difficulties are frequently brought on by the requirement to precisely calculate expenses, allocate them equitably, and maintain accountability and openness. Heterogeneous pricing, if not managed carefully, can lead to accusations of price discrimination, which may be illegal or unethical.

c. **Interoperability and Standardization:** There are numerous major challenges to heterogeneous interoperability and standardisation. These include organisational obstacles to sharing data and resources, security issues, lack of shared standards, and data heterogeneity (various formats, structures, and interpretations). Addressing data integration concerns, creating common standards, putting strong security measures in place, and encouraging cooperation across various organisations are all necessary to overcome these obstacles. Ensuring seamless interoperability between different technologies and vendors is crucial for maximizing the benefits of heterogeneous networks, and standardization efforts are needed to facilitate this.

d. **Security and Privacy:** The complex network structure of HetNet's, with its diverse base station (BS) types and complex network topologies, introduces significant privacy challenges. Insecure access points in this diverse environment jeopardise user privacy and data security and may jeopardise sensitive information. Heterogeneous networks can introduce new security and privacy vulnerabilities, requiring robust security measures to protect user data and network integrity.

1.3.1 Cost Optimization

LTE heterogeneous networks (HetNet's) can increase network capacity and coverage while potentially lowering costs by combining several base station types (macro, picot, femto, etc.) with differing power levels. Strategies to control interference, balance load, and lower operating costs are all part of optimising these HetNet's. Heterogeneous networks, provide options to reduce costs by selectively switching on/off nodes (cells) based on demand, hence lowering energy usage and operational expenses. There are several cost optimization strategies for Heterogeneous networks:

1.3.1.1 Cell Splitting

Reducing cell size (e.g., deploying pico or femto cells) increases frequency reuse and capacity, allowing for more efficient use of resources. In regions with high user density or coverage requirements, LTE heterogeneous cell splitting which entails installing a combination of macro, pico, and femto cells offers an affordable means of improving network capacity and coverage. This strategy minimises total deployment costs and optimises resource allocation by utilising the advantages of various cell types. Here is a breakdown of how LTE heterogeneous cell splitting achieves cost optimization:

1.3.1.2 Reduced Reliance on Macrocells

In cellular networks, "reduced reliance on macrocells" refers to strategies and innovations that decrease the need for conventional, sizable macrocell base stations to provide overall network capacity and coverage. Smaller cells, such as microcells and picocells, are frequently used in conjunction with other technologies, such as distributed antenna systems (DAS), to offload traffic and enhance user experience in particular regions. Macrocells, while providing broad coverage, can be expensive to deploy and maintain. By introducing smaller cells (pico, femto) to offload traffic from macrocells, operators can reduce the need for expanding macrocell coverage in densely populated areas, saving on infrastructure costs. This reduces the need for more macro cell sites, which are often difficult and expensive to acquire, especially in urban areas. The following cost benefit are reasons for reducing the overreliance on macrocells.

a. **Capacity limitations:** Despite their wide coverage, macrocells have capacity constraints, particularly in areas with high population densities, because of issues like bandwidth constraints and the maximum number of concurrent users they can accommodate. When numerous users are attempting to connect simultaneously lead to slower speeds and connection problems. Macrocells may cover a huge geographic region because they are made for wide-area coverage. However, capacity is frequently sacrificed for this broad coverage, particularly in places with a high user density. When numerous people attempt to access the network at once, the limited bandwidth available on these networks is shared, resulting in slower speeds. Although they can accommodate many users, the maximum number of users that can concurrently use the network without experiencing performance deterioration is limited. Once more, in densely populated locations, these networks may encounter interference from small cells or other macrocells, which could affect their performance and capacity.

b. **Coverage gaps:** Macrocells may not adequately cover all areas, especially indoors or in dense urban environments. Small cell deployment and the use of heterogeneous network (HetNet) architectures can help close LTE coverage gaps, especially in places where using macrocells alone is not enough. To increase capacity and coverage particularly in regions with high user density or signal interference these technologies deliberately position smaller, lower-power cells (pico, micro, or femto cells) within the coverage area of macrocells.

c. **Cost and energy consumption:** Numerous macrocells can be costly and energy-intensive to maintain and power. In LTE networks, less reliance on macrocells can result in more affordable costs and energy usage. This can be accomplished by carefully placing smaller cells such as femtocells or small cells to augment the coverage of macrocells, especially in high-density geographical areas. Because the macrocells can manage less traffic, this method lowers their power requirements and may result in more effective energy use overall. Because fewer macrocells are required when traffic is offloaded to smaller cells, equipment, installation, and maintenance expenses are reduced. Since small cells usually manage less traffic and cover smaller areas than macrocells, they can use less expensive backhaul technologies. More equitable traffic distribution allows operators to better allocate resources, preventing the need for costly macrocell infrastructure upgrades. Because macrocells offload to tiny cells, can run at lower power levels when they are not processing as much traffic. Small cells can use less power per unit of traffic managed than macrocells, especially those made with energy efficiency in mind. Energy efficiency can be further improved by designing small cells with sleep modes that enable them to shut down or use less power when traffic is light.

d. **Self-Organizing Networks (SON):** SON solutions automate operations like assigning physical cell identifiers and configuring radio resources, decreasing manual intervention and operational expenses. LTE Self-organising Networks (SON) can reduce reliance on macrocells by automating network configuration, optimisation, and healing, which includes the deployment of smaller cells such as picocells and femtocells. SON improves network performance, enhances user experience, and lowers operating costs. SON makes it easier to integrate small cells (pico, femto, and relay nodes) into existing macrocell networks, resulting in the formation of HetNet's. This densification with smaller cells relieves macrocells of the load while boosting overall network capacity and user experience. SON automates the initial setup and configuration of new cells, including PCI assignment and radio resource setting, making small cell deployment and management easier. The technology continuously analyses network performance and automatically adjusts parameters such as transmit power, handover thresholds, and resource allocation to improve coverage, capacity, and user experience, especially in locations where macrocells are overcrowded or have coverage gaps. SON has the potential of automatically detecting and resolving network connectivity problems such as interference, outages, and handover failures, reducing downtime, and ensuring network stability. This reduces the need for manual intervention and troubleshooting in macrocell-heavy networks. SON approaches such as fractional power control (FPC) reduce cell interference, which is especially critical in HetNet's with many tiny cells, ensuring effective resource utilisation while minimising negative affects on user experience. SON optimises handover procedures between different cell types (macro and small cells) to provide smooth user mobility and consistent service quality as users move around the network.

e. **Multi-Domain Load Balancing:** Balancing traffic among distinct domains (e.g., macro, pico, and femto) to maximise resource utilisation and minimise interference, particularly in overlapping areas. LTE multi-domain load balancing tries to improve network performance by distributing traffic over multiple network domains (e.g., macrocells, small cells) to avoid overload in certain locations and increase overall network efficiency. This technique decreases dependency on macrocells by using low-power nodes (LPNs) and other small cells to manage traffic, particularly in densely populated areas.

1.3.2 Network Ownership and Business Models

Ownership in heterogeneous networks refers to the person or organization that owns and manages the network's infrastructure and related resources. This can differ among various network components, resulting in intricate interactions between domains and difficulties in areas such as policy administration and security. To create successful business models, it is essential to comprehend the ownership structure of various network segments (such as operator-based versus operator-independent deployment). Business models in heterogeneous networks sometimes entail service providers selling services to MVNOs in bulk, with the MVNOs thereafter holding the customer relationship. This may resemble existing mobile network operator (MNO) arrangements whereby MVNOs purchase services from terrestrial networks in bulk. Network as a Service (NaaS) and roaming are two examples of services that alternative access network providers can offer as wholesale operators.

Heterogeneous networking gives users a more dependable and seamless experience since they can shift between diverse types of wireless networks when they travel from one location to another. This can result in enhanced user experience and greater levels of satisfaction. While energy savings and cost savings are crucial, it is essential to ensure that the implementation of heterogeneous networks has no negative impact on user throughput or overall network performance. Furthermore, social, and economic constraints can impact the establishment of diverse networks, influencing the variety of user behaviours and economic outcomes.

1.3.3 Total Cost of Ownership

Total Cost of Ownership (TCO) in heterogeneous networks refers to the sum of all direct and indirect costs associated with acquiring, deploying, operating, and maintaining a network over its entire lifecycle. This includes expenses like initial hardware, software, and implementation costs, as well as ongoing maintenance, support, and operational expenses.

For a mobile wireless network operator, capital expenditure (CAPEX) for network building and the operating expenditure (OPEX) for network operation are the most crucial components of the total cost of ownership (TCO). A conventional wireless network's cost structure is comparatively well known [5]. The cost of the mobile core network (the gateways and IMS platforms), the radio access network (the base stations and radio network controllers), the backhaul infrastructure, and site acquisition, construction, engineering, and integration are typically included in the CAPEX for a typical cellular network.

About 60% of CAPEX goes toward the radio access network, 15% goes toward the core network, 5% goes toward backhaul, and 10% to 20% goes toward site acquisition, construction, and engineering. Of course, there are significant differences between one deployment to the next. For instance, if an operator can upgrade its current base station equipment to support a new air interface technology, the cost of the radio access network can be decreased; if an operator can overlay a new network on its existing cell sites, the cost of site acquisition can be avoided.

OPEX is mostly related to network management and operation, which includes site leasing costs, backhaul transmission costs, network operation and maintenance costs, and electricity costs. OPEX make up as much as 60% of the TCO because base station equipment depreciates over a 7-year period [3]. Typically, neither CAPEX nor OPEX includes the cost of obtaining wireless licenses. This is because these kinds of purchases are rare and occasionally very costly, costing anywhere from several billions to tens of billions of dollars. The spectrum cost may account for a sizeable amount of the TCO in some markets. A heterogeneous network's cost structure is frequently quite different. In a heterogeneous deployment, there are typically many small cells, even though their costs are typically an order of magnitude lower than those of a macrocell base station. The backhaul cost may rise dramatically in tandem with the number of cell sites.

The cost structure for site acquisition, installation, and site leasing may be different since many small cells are installed indoors on the wall or outdoors on utility poles. For many small cells, the running and maintenance expenses also provide another challenge for the heterogeneous network operators. Techniques for self-organizing and self-optimizing become indispensable to lower the general network cost. Femtocells installed by consumers could complicate the matter even more. In this case, a mobile operator may pay a much higher cost in customer technical support even if site acquisition, equipment installation, and back haul connectivity may cost little or nothing. A heterogeneous network is potentially made up of network layers that operate on unlicensed bands, such as Wi-Fi.

While unlicensed bands do not incur a licensing cost as a conventional cellular network, the uncontrolled radio environment may be difficult to manage and run, especially in very dense hotspot installations. The effect of cost structure on heterogeneous networks has been investigated in several works. [6, 7] presents a good example since it suggested a technique for evaluating the total cost and performance of a heterogeneous network made of several base station classes and radio access technologies with various cost and technical characteristics.

It also shows the effects of the change in business model and cost structure. Another excellent illustration comes from [3], which emphasizes how roughly 41% of OPEX annually is spent on electricity at macrocell sites—of which 48% is consumed for air conditioning. Using advanced base station design and network architectures including fan-less cooling small cells and centralized baseband processing of large-scale cloud base stations helps lower the energy consumption as the industry keeps pushing for higher energy efficiency. Regarding TCO for the mobile wireless sector, the development towards heterogeneous networks offers both possibilities and difficulties.

This is caused in part by the substantial change in the usage model from outdoor to indoor/hotspot, from high system capacity to both high system capacity and high data rate, and in part by the substantial change in the cost structure of base stations, backhaul, network installation and maintenance, energy costs, and radio spectrum costs.

1.4 Cellular Heterogeneous Networks in Application

To satisfy various market demands and financial constraints, heterogeneous networks come in a wide variety of network architectural designs and implementation variations [8]. The objectives are comparable, Heterogeneous networks must offer consumers 24/7 coverage, high data rates, high capacity, security, and the best possible network user experience. Fast time-to-market, optimal network utilization, operator control, and network manageability are all requirements for mobile operators in heterogeneous networks.

Home femtocells are the most traditional deployment of a heterogeneous network. They are mostly used by end users to extend network coverage in indoor spaces or isolated rural areas where outdoor macrocells struggle to provide coverage. They are mostly utilized for voice over IP services and data as an extension of cellular coverage. Since the consumer's own internet service, these home femtocells are connected to mobile core network, the mobile operator typically incurs no backhaul transmission costs.

Since the user bears the cost of the backhaul transmission, a home femtocell frequently restricts access to a "closed subscriber group." This makes financial sense to the user, who has little incentive to share it with others. Most home femtocells are installed by consumers who may not possess sufficient knowledge of radio technologies, which is one of their distinctive features. Therefore, it could be challenging to control RF interference from or to a home femtocell. Assuming the mobile operator has enough spectrum for a stand-alone RF carrier, a common solution is to assign home femtocells to a different carrier frequency than the macrocellular network.

The so-called picocell, also known as an enterprise femtocell or metro femtocell, is another type of small cell. They function as an extension of the macrocellular network and employ the same air interface technology across licensed spectrum used by mobile operators. Compared to home femtocells, this class of small cells typically has a larger subscriber capacity. It offers voice and data services in office settings, indoor coverage in locations like shopping malls, and outdoor hotspot coverage in places like a busy shopping street or sports stadium. They are frequently professionally installed with more sophisticated antennas, environmentally hardened, especially for outdoor deployment, and offer service to all eligible subscribers rather than just those in a closed subscriber group.

It should be mentioned that the subscriber capacity and transmission power of home femtocells and picocells might not differ significantly. The primary distinctions lie in the way they are linked to the core network of the mobile operator, a crucial topic that we will cover in a later section. Because picocells frequently require carrier-grade equipment, their average cost ranges from one to several thousand dollars. Data traffic offloading from cellular networks to Wi-Fi has increased significantly in the last few years.

Both the licensed (for 2G/3G/4G cellular) and unlicensed (for Wi-Fi) bands are used by this kind of heterogeneous network. The integration of Wi-Fi and cellular networks necessitates careful economic consideration and varies from market to market because cellular networks and Wi-Fi differ in terms of user/device credentials, authentication, air interface characteristics, network architectures, interference environments, and subscriber management and billing systems. While some mobile operators opt to set up their own Wi-Fi networks, others use third-party Wi-Fi service providers to deliver services under their own or a third-party brand.

Additionally, the scale of the Wi-Fi networks vary, ranging from a few thousand access points (APs) for airport hotspot coverage to over two million APs for large city coverage nationwide. The pricing structure is often complicated because of the vast differences in who owns network equipment, who provides backhaul transmission, and how cell site real estate is arranged.

Table 1.1 Home femto Vs Public Picocell Key Features

Parameter	Femtocell	Picocell
Deployment	Unplanned, Consumer deployment	Planned, Operator deployment
Installation	Consumer installation	Operator installation
Equipment	Consumer Owned	Operator Owned
Enclosure	Consumer electronics	Environmentally hardened
User access	Closed User Group, restricted access	Open Use Group, to all qualified subscribers

Parameter	Femtocell	Picocell
Backhaul	Consumer grade, paid for by user	Carrier grade, paid for by operator
TX power	<250 mW	250 mW to 2 W
Capacity	<8 users	16–32 users
Handover	Loose coupling at network layer	Tight coupling, intra-network handover
Security	Untrusted by operator network	Trusted by operator network

The precise data offloading strategy further complicates the business model. While some operators opt to offload only the radio link and send the traffic to the mobile operator's core network via Wi-Fi, others opt to offload both the core networks and the radio access, sending the Wi-Fi traffic straight to the public internet.

1.4.1 Heterogeneous Network Cohesiveness

To create a cohesive and effective network infrastructure, HetNet, or heterogeneous network, refers to the integration of multiple network types, such as Wi-Fi, in-building wireless systems, radio access networks, and macro sites. The Deployment of Heterogeneous Network usually focusses on extremely high-loading environments, like a sports stadium. The convergence of Hotspot 2.0 and Access Network Discovery and Selection Function (ANDSF) for network interoperability, the shift from providing coverage or capacity to offering value-added services by integrating location and proximity services into the integrated small cells. The potential for creating super control nodes that coordinate radio resources across multiple network layers are some of the general trends in the development of heterogeneous networks.

The contention between Wi-Fi access points and femto stations is mostly resolved. Their complimentary nature is now widely acknowledged. In order to achieve its economic potential, a practical heterogeneous network must ultimately provide three things: technically, it must close the capacity gap caused by the data tsunami; commercially, it must assist mobile operators in generating new revenue streams and differentiating their services while minimizing CAPEX and OPEX; and for end users, it must provide an exceptional user experience through always-on and always-best-connected. To accomplish these objectives, a heterogeneous network solution must make use of current deployment and technology, enhance and grow current applications, and facilitate upcoming service developments.

1.5 Heterogeneous Network Technology Overview

This section covers the various technical facets of heterogeneous network technology, including its difficulties, trade-offs, and potential future developments.

1.5.1 RF Interference

When 3G/4G femtocells are put inside a 3G/4G macrocell network's service region, the two network layers may use the same carrier frequency or two different carrier frequencies. The benefit of sharing a frequency is that it uses the least amount of spectrum, which is crucial for operators with a limited supply of spectrum. However, if these two networks are not correctly synchronized in design, there may be significant interference between them.

A UE connected to a macro base station, for instance, may cause excessive interference to neighboring femtocells, especially when the UE is transmitting at the cell edge near its maximum power level. Desensitizing the femto station receiver to prevent RF saturation is a workable way to deal with this issue. The femto stations should be placed in areas that offer some inherent RF signal isolation from the macro network, and the handover triggers should be carefully designed to ensure dependable UE mobility between these two networks.

Setting up macro and femto stations on different carrier frequencies still causes UEs to struggle to free from interference. The pilot signal from the macro station can overwhelm the weaker pilot signal from a femto station, even if the UE is already well within the femto station's coverage area. This is usually not the case in heterogeneous deployments. However, the industry adopted the "range extension" technique, which basically adds a bias value to compensate for the weaker femto pilot and thus "extend" the coverage area of a femto station, because such a change would necessitate some fundamental modifications to the deployed equipment and the current standards [9].

Another obstacle to managing RF interference in real consumer femtocell deployment is the closed-subscriber-group (CSG). Access to the femto station is limited to a specific user group, typically family members or house guests, because customers typically use their own internet subscription for the femtocell backhaul connection. Even if a consumer-deployed femto station is nearby, a UE cannot reach it unless it is a part of the CSG. In a crowded femtocell deployment context, like an apartment building, this can cause an excessive level of interference to the networks. Since all users often have access to every femto station in the office or home, this is fortunately less of an issue for enterprise settings. Femtocells' open subscriber group has been discussed; however, this is more of a business model problem than a technical fix.

Wi-Fi networks present a vastly different scenario. Although there is not a "macro" Wi-Fi station to be concerned about, the unlicensed-band nature of the network makes interference between Wi-Fi access points and devices more unpredictable. Multiple Wi-Fi networks are serving an increasing number of hot spot regions as capacity demand rises, which might lead to severe interference between them. The wireless industry is examining new deployment in the upper unlicensed spectrum, such as the 5 GHz band, local Wi-Fi channel coordination, and network sharing (e.g., many service providers sharing one access point by broadcasting multiple SSIDs over the air). It is also anticipated that future heterogeneous networks will accommodate novel network architectures and forms of connectivity.

Innovations of the future include D2D (device-to-device direct communication [10]), mobile relay (e.g., a relay station installed in a high speed train to provide relay service between the passengers in the train and the base station network along the trackside, layer 2 and layer 3 relay (including in-band backhaul applications similar to a mess network), traffic aggregation point for machine-type communications (a small radio station that provides shortrange connectivity such as ZigBee to local sensors and relays the aggregated traffic to the network through a long-range cellular connection), and cognitive radio for spectrum sharing (e.g. in TV white space bands).

A heterogeneous network RF engineer must consider a common set of requirements, such as frequency planning, co-existence, transmission power level, and receiver sensitivity, both within a heterogeneous network and within a neighboring homogeneous or heterogeneous network run by a different mobile network operator, even though the RF interference issues and challenges vary greatly from case to case. Currently, most studies on heterogeneous networks concentrate on creating solutions for a particular set of network layer combinations. To scale up the design and deployment of heterogeneous networks through an optimal distribution of functions among devices, access points, and the cloud, a longer-term vision and framework are required.

1.5.2 Heterogeneous network environment

Conventional cellular networks adhere to stringent network design guidelines, and each cell site is chosen separately and strategically. Real-time channel quality, interference level, and resource requests are some of the factors put in design; however, the network decides how best to manage radio resources. In homogeneous networks, network-centric design has proven successful in striking the ideal balance between capacity and coverage. Traditionally, cellular networks were developed as homogeneous networks, which meant they were made up of base stations with comparable features (transmit power, antenna layouts, etc.) and provided consistent coverage. However, the growing demand for data and diverse user needs have resulted in the evolution of heterogeneous networks (HetNets), which combine macrocells and smaller cells (microcells, picocells, and so on) to improve capacity and coverage. More diverse settings, including homes, workplaces, retail malls, and sports stadiums, both indoors and out, are used for the deployment of small cells.

As cloud-RAN and small cells proliferate, SON technology continue to advance. The ability of SON technology to optimize across various network tiers and base stations within a given geographic area is a solemn idea worth its salt. We anticipate a greater role for mobile devices. For optimal performance in a heterogeneous network environment, we anticipate that SON will make extensive use of cloud-RAN technology to enable each network layer to configure its radio system and shift more decision-making to devices, one potential solution is to have each network layer customize its radio system and transfer additional decision-making authority to devices by providing policies and broad information about the RF environment.

1.5.3 Network Coupling

The 3GPP standard [11] states that an "enhanced packet data gateway" (ePDG), which offers real-time packet processing, subscribers, service, and application monitoring, and backhaul security via IPSec tunnels, can be used to connect an "un-trusted" Wi-Fi network to a 3G cellular network. However, The Enhanced Packet Data Gateway (ePDG) is equally a critical network component in 5G and 4G LTE for secure access and interoperability between 3GPP networks (such as LTE and 5G core) and non-3GPP networks (such as Wi-Fi). It creates a secure tunnel (IPSec) for user data, allowing seamless connectivity and traffic offload between Wi-Fi and cellular networks.

Loose Coupling: When the 3GPP core network and the Wi-Fi access network share only the AAA server for authentication, we regard these two networks as being weakly linked. Each network functions autonomously in this loose-coupling scenario, and neither network needs to modify its protocol stack or design. In traditional interworking (Loose Coupling), the LTE and Wi-Fi networks are independent, with a gateway (e.g., Packet Data Network Gateway or PGW) acting as the interface. Herein, the gateway adds latency and complexity to data flow between networks.

Tight-coupling: In a tight-coupling scenario, Wi-Fi data travels in the same manner as cellular data from ePDG to the 3GPP packet data network gateway (PDN-GW) via the 3GPP core network. Wi-Fi users can access 3GPP services with the possible support of guaranteed QoS and seamless mobility in addition to 3GPP authentication, as in the loose-coupling scenario. Tight coupling can also enable other 3GPP services for Wi-Fi users. A heterogeneous network can establish a tight coupling at lower layers between two radio access networks in addition to the tight coupling at the core network level that we just discussed. This would enable one radio access network to have specific visibility and control over the other network's radio resource management. One pertinent example is the CDMA-LTE changeover design, which may provide the same degree of quick handover performance as a homogeneous network through tight coupling at the radio link layer.

Tighter coupling: Tighter coupling, as used in LTE and Wi-Fi integration, is a means of merging the two technologies at the radio protocol stack level, allowing for more seamless and efficient data transmission between them. This arrangement, also known as "very tight coupling," differs from standard methods in which LTE and Wi-Fi operate as independent networks with a gateway in between. Tighter coupling (very tight coupling): In Tighter coupling the LTE and Wi-Fi are merged at the radio protocol stack, typically at the Packet Data Convergence Protocol (PDCP) layer. This creates a unified network experience, allowing users to access both with a single IP address. Utilising both technologies concurrently allows for faster network switching and increased data transfer speeds. LTE Wi-Fi Radio Level Integration with IPSec Tunnel (LWIP) is one design that can help achieve this goal. LWIP enables dynamic traffic steering between LTE and Wi-Fi networks based on link quality.

Table 1.2 Network Coupling Types

Coupling type	Characteristics	Level of Complexity
No Coupling	Two Separate networks operate independently.	Here there is no change to existing networks, suitable for roaming architecture. MNO may perform offline billings consolidation for single bill.
Loose Coupling	Two networks share user credentials and AAA. data traffic goes through separate core networks	Common user/device credential for two networks. Eliminate user intervention for Wi-Fi access to core network and for traffic offload to internet

Coupling type	Characteristics	Level of Complexity
Tight Coupling	Tight coupling in the context of femtocells refers to a network architecture where femtocell base stations are tightly integrated with the macrocellular network, often at radio or core network levels. This integration can involve directly connecting femtocells to the macrocell network's evolved packet core (EPC), as standardized by 3GPP in Release 10.	Opportunity for operators to offer value-added services.
Very Tight Coupling	Two access networks are directly interconnected with real-time radio resource management across networks, (e.g., picocell connected through X2 Application Part interface)	This is an opportunity to implement carrier aggregation and interference coordination which requires significant design and implement efforts

In application If a cellular operator uses a third-party Wi-Fi network to establish a heterogeneous network, loose coupling may make both networks easier to operate in terms of business ties. Operators that own both networks but do not wish to make major alterations to the equipment that is currently in place may also be in this situation. The use scenarios of an operator also play a significant role in the decision regarding network coupling. To avoid core network congestion, some operators may decide to use the networks for different purposes, such as one for voice calls and smartphone data access and another for data offloading to the internet. In this case, the benefits of tight coupling and high-performance service continuity may take a back seat. To give operators a wide variety of technological options, the industry tends toward tightly integrated solutions with secure data offloading and service continuity from the perspective of standards development. These solutions continue to evolve for use in a range of industries, including residential, commercial, and cellular carriers.

1.5.4 User Equipment/Device Authorization Credential

The user or device credential used for accounting, authorization, and authentication is a crucial technological component in a heterogeneous network. Mobile phones' SIM cards, laptop computers' certificates, embedded and soft SIMs in certain machine-type communication devices, and Wi-Fi login credentials are the most widely used user-device credentials. The legacy of the information technology and telecom sectors was primarily responsible for these disjointed user-device credential solutions, which are now posing a growing obstacle to the deployment of heterogeneous networks. For instance, most Wi-Fi-only devices now utilize certificate or username-password combination for access authentication, whereas most mobile handsets use SIM cards. For smooth network access, enhanced user privacy and information protection, uniform service availability, consistent quality of experience, and easier device administration, a unified user-device credential solution is ideal.

Contritely, the necessary modifications to the design of the portable device platform and, more importantly for operators; the accounting and invoicing system may take some time to implement. For seamless network access, enhanced user privacy and information security, consistent service availability, a consistent quality of experience, and streamlined device administration, a unified user-device credential solution is recommended. We expect solutions to converge gradually or not at all in circumstances with loose coupling. The growth of machine-type communication devices further complicates this problem. Some machine-type communication devices require a significantly longer life cycle (seven or more years) than consumer electronics like smartphones. They also need to be able to withstand deployment with the least amount of physical protection and be environmentally hardened. As a result, new types of device credentials can be needed. The wireless industry has, however, taken action to resolve the problems, and a few projects are presently in various phases of development [12]. The process might be accelerated by the confluence of information technology and telecommunications.

1.5.5 Interworking

The process of integrating various wireless network types, such as Wi-Fi and cellular networks, to enable seamless data services and communication is known as cellular network interworking. By utilizing the positive aspects of each network, this integration enables users to move between technologies according to their location and bandwidth requirements. Managing mobility, security, and quality of service (QoS) across several network types is part of interworking. Interworking provides a certain amount of efficiency for multi-mode devices that need movement across different networks. Interworking mobility entails two approaches. The first is roaming, which occurs between geographically dispersed networks. The second is between co-located networks, like handover between several network levels in a heterogeneous network. Support for roaming and handover is one of the essential features that allow end users to easily access essential services on heterogeneous networks and the multi-mode devices that are connected to them.

The access gateway or an interworking function node are frequently used to accomplish interworking between networks. Such nodes offer intermediate protocol termination on data planes and protocol conversion or translation on the control plane, as various network layers may have distinct network architectures and protocol stack designs. In situations where two networks have different owners, this node may also act as the security boundary between them.

The access gateway equally provides security and an interworking function for handovers, the home NodeB gateway specified in 3GPP establishes an IPsec tunnel to the home NodeB (sometimes referred to as a femto station) over the public internet. An interworking node, in this case the ePDG, gives Wi-Fi devices access to the AAA server in the 3GPP core network for shared authentication and charging because some networks, like Wi-Fi, lack the core network architecture and related protocol stack. To give end users a more seamless experience, future developments will also include unified QoS capability across various network layers.

The primary obstacle to interworking is that when the number of layers in a heterogeneous network are increased, in such a case, the complexity of both design and implementation increases. It takes a lot of engineering work to get all the layers to communicate with one another because each access technique often has its own control plane design, especially for tight coupling. With the advent of novel network topologies (such as mobile relay and device-to-device communications) and connectivity types (such as ZigBee for machine-type devices), we anticipate that the difficulty will only grow in the future.

To solve the problem of interworking complexity, the wireless industry is considering two technological avenues. One option is to give mobile devices additional intelligence, which would allow them to manage some of the network infrastructure's interworking complexity. Creating and implementing advanced connection manager software for mobile devices is one example. The second is to transfer some of the complexity from the control plane to the data plane to enhance data plane processing. Improved data plane processing is especially appealing for applications that need consistent QoS compliance across layers of a heterogeneous network since the device has comprehensive knowledge of the apps it is running.

1.5.6 Handover

The movement of a mobile device outside the network's service area can be effortlessly relayed from one cell or base station to another. This is referred to as traffic handover in the telecom industry. This process ensures that communication continues by connecting to a new base station without interfering with the current session. Handover performance and reliability are critical performance measures in heterogeneous networks, just as they are in homogeneous networks, and they have a direct impact on user experience.

In a homogeneous network, handover is mostly caused by a mobile device departing the coverage area of a serving base station or network. On occasion, a wireless operator may opt to instruct a device to switch to a specific base station to maintain the network or balance the load. A high-power macro base station network often provides high-speed mobility assistance and ubiquitous network coverage in a heterogeneous network, but low-power micro cells allow for more cost-effective data offload at hotspot zones.

The methods and processes for intra-system handovers in the same layer of a heterogeneous network are like those in a homogeneous network. When it comes to inter-system handovers spanning multiple layers, the decision to hand over becomes significantly more complex. Many decisions in a heterogeneous network are based on factors such as the cost-of-service delivery to an operator, data pricing to users, the QoS requirements of specific applications, and the non-ubiquitous nature of the hotspot network layer, in addition to the traditional coverage-reason-triggered handovers. For example, as soon as a mobile device reaches the hotspot or home network coverage area, it is best to switch it from the macro base station network to small cells for data offloading.

The degree of network connectivity has a considerable influence on the design of the handover algorithm. Because of service continuity and quality requirements, it is preferred to avoid inter-layer handover for conversational and real-time applications such as phone calls. Because of the reduced service interruption, a heterogeneous network with tight coupling between layers can withstand more aggressive inter-layer handover decisions than a comparable network with loose coupling. As a result, unlike homogeneous network, designing the handover method in a heterogeneous network environment is far more difficult and complex.

Five steps make up a handover process in heterogeneous networks, which is comparable to the conventional designs that are frequently employed in homogeneous networks: (1) handover preparation, (2) handover trigger, (3) handover decision, (4) handover execution and session transfer, and (5) handover completion. The mobile device searches for potential handover targets based on pre-optimized mobile-network association criteria as part of the handover preparation process. The serving network may provide specific information about the candidate networks, such as the type of air interface technology and the basic radio configuration information, including carrier frequency, to help with this search process and reduce the mobile device's battery consumption. To maximize performance, the providing network could also give the mobile device specific handover-related parameters.

A mobile device may also pre-register with the target networks to save time during authentication and service authorization when handover happens. A heterogeneous network must, however, pay a price for all these capabilities, such as ensuring uniform registration from a mobile device and demanding that multiple network levels have appropriate information about one another. This complexity grows rapidly when the number of layers increases or the deployment environment gets more complex (for example, a mix of indoor and outdoor networks or various network topologies).

Fortunately, researchers from academics and business are creating advanced ways for network self-configuration. Device pre-registration administration may become easier because of efforts to standardize user and device credentials. In a homogeneous network, inappropriate handover triggers result in wasteful use of radio resources and/or radio connection failure (such as call dropouts). Knowing where and when to transition between networks affects user experience and network performance in a heterogeneous network, as small-cell coverage may not be ubiquitous.

The transition trigger algorithm for a heterogeneous network must include operator policy, subscriber use cost, and quality of service needs, as well as radio link conditions, making the design even more complex. These demands may vary depending on the time of day and/or geography. Following activation, an entity in the system must determine whether to hand over. In traditional cellular system designs, the infrastructure network is responsible for determining the final handover point. The mobile device is responsible for submitting a handover request to the network as soon as the handover requirements are met, however the network can either execute a handover instruction or do nothing.

In heterogeneous network situations, this network-centric handover decision-making process has limitations. For example, the heterogeneous network's serving layer may lack all the essential information or capabilities to determine which layers to transfer control to. This is especially true for networks that are loosely connected. In future growing heterogeneous networks, we expect mobile devices to play a more active role in handover decision-making. As a result, some handover communication flows will change.

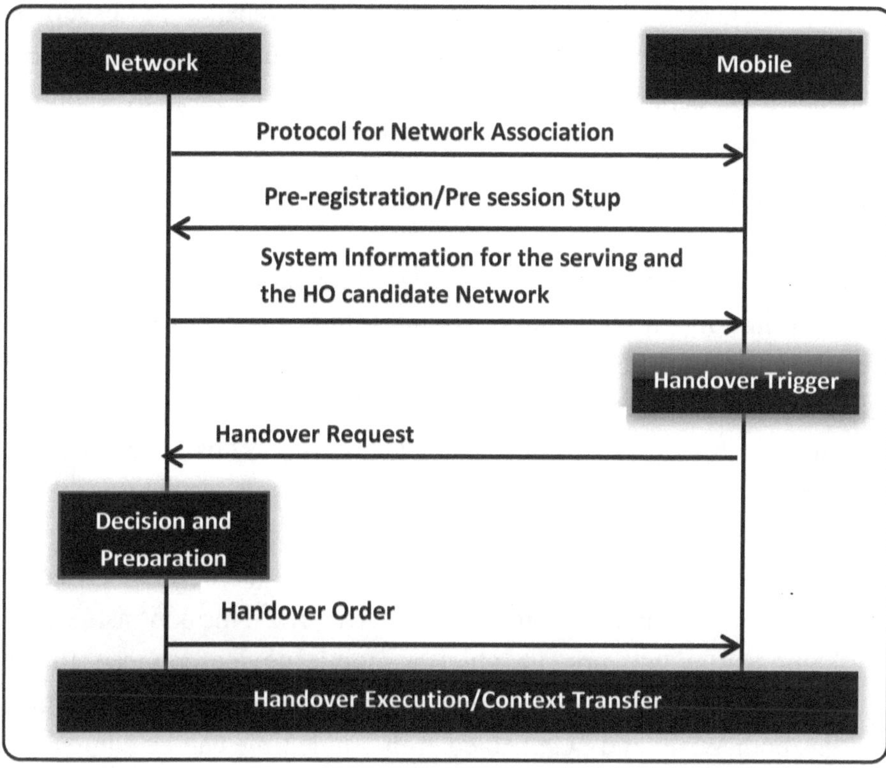

Figure 1.3 An example of a mobile-triggered network-decided generic handover process.

The same strategies used in homogeneous networks can be applied to heterogeneous networks to reduce service disruption and execute handovers. As with traditional handovers, the source network may instruct the destination network or networks to pre-fetch the necessary context data before the actual handover. During handover execution, session information, including the remaining user data in the transmit buffer, may be shifted from the source network to the destination network, depending on the extent of network coupling [13]. Following the handover procedure, the heterogeneous network requires minor maintenance to sustain its overall software robustness and resource utilization efficiency. Figure 1.3 depicts a typical handover method initiated by a mobile device and determined by the network.

1.5.7 Data Routing

Data packet routing is a network layer function, whereas handover is primarily a link layer issue. Data packet routing in a heterogeneous network consists of two basic components: the selection and/or aggregation of radio links over the air and the IP packet procedure in the network. Data routing is designed with both technical and business considerations in mind. A heterogeneous network allows different air interface technologies to operate concurrently, potentially resulting in multiple radio links between a mobile device and the heterogeneous network. This contrasts with a homogenous network, in which a mobile device connects to the network using a single radio link. A smartphone, for example, which has both Wi-Fi and 3G enabled at the same time can be aggregated to use available capacity for the increase of data rate.

The alternative is to intelligently map user application flows to various radio links based on user preferences, operator policies, and each link's specific QoS requirements. For example, the mobile device can use the Wi-Fi link for high-speed best-effort internet access and the cellular connection for voice calls. This functionality is supported by 3GPP's IFOM (IP flow mobility) feature. A heterogeneous network may provide many routing possibilities within the infrastructure network to various levels, in addition to radio link selection and aggregation over the air. This adaptability becomes increasingly important as the industry faces an increasing number of core network overloading issues as air interface capacity rapidly grows due to the deployment of small cells and more spectrally efficient air interfaces such as LTE and LTE-Advanced.

To offload an operator's core network or to selectively route traffic to the public internet or to the core network based on operator business considerations, 3GPP is developing several solutions, including SIPTO (Selective IP Traffic Offloading) and LIPA (Local IP Access, also known as local breakout for 3GPP femto stations) .In telecoms industry, there is a lot of research and technology development cantered on the coupling of multi-radio-link capacity with network routing flexibility. This is one of the unique instruments for implementing an operator's data offloading plan and achieving the greatest possible balance of service revenue and user experience in a diverse network.

1.5.8 Quality of Service

Any wireless network must emphasize providing end users with stable high-quality service. It is one of the most essential technological components of future wireless networks since it allows network operators to differentiate their services and earn more revenue. Even in a traditional homogeneous network, ensuring quality of service is challenging and complicated since it is end-to-end. Primarily, the network must be capable of creating and administering service policies, as well as having the necessary pricing and accounting architecture. When a subscriber requests a service, the network verifies the user based on his or her subscription record. The network then configures each network node along the availed data.

When network nodes start executing the desired QoS treatment for a particular application IP flow, the network must be able to track the actual service quality supplied to the end user. When the application session is completed, the network must collect all relevant accounting data and perform system inhouse keeping. This complex process encounters a variety of challenges especially where different networks are coupled. To begin with, a wireless network only has control over the radio link and the operator's core network; quality of service is meaningless to a user unless it is delivered end to end, the control that a wireless operator loses when the network is coupled. This has not been a major issue because the radio link has always been the service quality bottleneck.

However, in recent years, the proliferation of video and other media-rich contents has begun to cause congestion in backbone networks.

The second issue is that QoS implementation is not consistent. Different network segments may use various parameters in traffic scheduling algorithms and interpret QoS labels differently. This is particularly challenging when traffic goes over networks managed by many operators. The third issue is that users may not always receive the desired outcome from QoS scheduling. In addition, many popular applications and content are "over-the-top" traffic, which means they are supplied directly from the provider to the client via an open Internet connection. "Over-the-top" traffic, often known as OTT traffic, is multimedia content (such as video, audio, and other media) provided directly to consumers via the internet, bypassing traditional broadcast, cellular network, cable, or satellite TV services. The "over" component of "over-the-top" alludes to going beyond or beyond standard delivery techniques.

Support for QoS in heterogeneous networks is more challenging. The various network levels involved may use different network designs, air interface technologies, and QoS methods. Backhaul transport across the public internet for femto stations may have an impact on QoS management and delivery, even in heterogeneous networks using the same air interface technology, such as 3G/4G macro stations and 3G/4G femto stations. Some installations might be unable to provide consistent quality of service (QoS) across weakly connected networks due to extra variances caused by network coupling.

In heterogeneous networks with data routing capabilities between layers, the routing change itself may introduce additional latency into the data stream. Finally, the status of RF interference in unlicensed bands is frequently unexpected and changes over time, particularly in congested hotspots where QoS is most critical. As previously stated, wireless operators must make significant investments to fulfill the capacity demand caused by the surge in mobile internet traffic, even though their traditional voice and SMS service revenues are dropping.

As a result, quality of service (QoS) is becoming increasingly vital in maintaining revenue stream and standing out in a highly competitive marketplace. In the coming years, we expect to see more research and development on QoS for heterogeneous networks. Today, a heterogeneous network serves as a best-effort data conduit because a sizable portion of mobile internet traffic is over-the-top and originates from sources other than the operator's network, such as YouTube, TikTok and Skype.

The 3GPP's MOSAP refers to the work item "Interworking between Mobile Operators using the Evolved Packet System" within the 3rd Generation Partnership Project (3GPP). It focusses on ensuring seamless communication and service continuity when mobile carriers switch between multiple access systems, including both 3GPP (such as LTE) and non-3GPP networks. As a framework, 3GPP's MOSAP development (interface between data application providers and mobile operators using the evolved packet system) can be used. We call this a control-plane centric strategy because it requires the interchange of application needs and policy information between the data application provider system and the mobile operator network's control entities. This enables the mobile operator to properly arrange the data pipeline for QoS support. However, many data application providers may be unwilling to spend further in this regard and negotiate a business arrangement with mobile network operators because many over-the-top applications are already installed and operating effectively in the field.

Furthermore, scaling up the solution may make interoperability testing challenging. Finally, there is the challenge of managing QoS in a diverse network. Adding sophisticated signal processing to the data plane and transferring some complexity from the control plane to the data plane is a whole different strategy. Deep packet inspection (DPI) on both the mobile device and base station networks is a well-known example. When a radio link is transferred between different network layers in a heterogeneous network, the data plane's independent QoS handling capabilities is incorporated via DPI. To address network overloading scenarios, data plane packet processing can be integrated with other innovative content distribution technologies for the base station network, such as caching, cross-layer optimization, and adaptive media reformatting. As telecom and information technologies converge in cloud networks, data plane processing becomes increasingly practical and enticing. Moving complexity from the control plane to the data plane for mobile devices may have the advantage of reducing signal processing time.

Mobile devices benefit from understanding of their RF environment and the availability of each network in a heterogeneous network. It makes sense that mobile devices will play an increasingly significant role in decision-making and quality of service management. In the coming years, we expect the academic community and the wireless sector to continue to investigate traditional QoS methodologies and seek innovative technology opportunities in the context of media content distribution for heterogeneous networks and the convergence of telecom and IT. We believe that multi-communication devices and cloud network design will be critical factors in ensuring QoS across heterogeneous networks. It is also one of the most important economic aspects for mobile network operators since it may provide end consumers with an unequalled level of service quality and consistency in an extraordinarily complex and sometimes baffling heterogeneous network environment. It may also create a big income potential for operators striving to become featureless transparent-data-pipe utility providers.

1.5.9 Heterogeneous Network Security and Privacy

Network applications for machine-type linked devices will become more prevalent in the future information society because of security and device interconnectivity incorporated into a heterogeneous network and inside each network layer's protocol stack, ensuring a consistent and engaging user experience. Customers expect access to applications and data from anywhere as smartphones, tablets, and smart TVs become increasingly common, yet each technology has distinct security challenges, resulting in unequal user experiences. Security is essential for machine-type communications such as smart grids and smart transportation because it protects key infrastructure, ensures the seamless operation of an intelligent society, and secures the future embedded infrastructure.

Security concerns continue to grow as mobile internet becomes more incorporated into people's daily lives and a part of our modern culture. Any serious security breach might affect many users globally, as smartphone operating systems are increasingly becoming prevalent to a global population.

The desire for enhanced user convenience needs additional security procedures in addition to the broad possible security mitigations. Unified billing and single-sign-on (SSO) are fully dependent on secure connectivity between different systems. The Unified billing and single sign-on (SSO) are distinct but related concepts in IT. Unified billing simplifies customer billing by consolidating payments for multiple services into a single invoice. It makes it easier for users to access numerous apps using a single set of login credentials. Convenience is pointless unless it is safe. One solution being attempted by the industry to improve security is the direct embedding of security elements into silicon chips. Concerns or understanding of privacy have grown along with the popularity of mobile internet. Protecting privacy is even more important in heterogeneous network environments because layers of the network must transmit key information to maintain the computing continuum, between various operators that control and run each layer of the network.

The collection and, more importantly, distribution of user and application data may become increasingly important as mobile network operators strive to provide context-aware and location-based services rather than a featureless data stream. Device-to-device (D2D) communication eliminates the requirement for a central network infrastructure by allowing direct communication between devices that are close to one another. Benefits of this technology include lower latency, higher data throughput, and enhanced spectrum and energy efficiency. D2D communication is frequently used in situations like file sharing between cell phones or connecting Internet of Things devices. It can make use of a variety of wireless technologies, including Bluetooth, Wi-Fi Direct, and NFC. Using device-to-device (D2D) for social networking apps as an example, this functionality will never be useful if everyone stops it due to security and privacy concerns. Privacy has always been a concern for location-based services, particularly in a heterogeneous network environment with several network owners. We expect mobile network providers taking on more responsibility for regulating and maintaining security and privacy, as well as relying heavily on subscriber ownership and cloud infrastructure.

1.5.10 Capacity and Coverage Enhancement Measurement Challenges

LTE capacity and coverage enhancement metrics encounter numerous obstacles, including the difficulty of optimising for both capacity and coverage at the same time, the impact of inter-cell interference, and the fluctuation of channel conditions and user behaviour. Furthermore, balancing network performance with the expense of adding and operating more cells complicates matters. Challenges for Improving LTE Capacity and Coverage are:

a. **Conflicting Optimization Goals:** Maximising both capacity and coverage in an LTE network is an arduous task because optimising one may have a detrimental influence on the other.
b. **Inter-Cell Interference:** Inter-cell interference can have a major impact on network performance, especially in locations with high cell density. Managing this interference necessitates careful coordination among neighbouring cells.
c. **Channel Variability:** Signal propagation in LTE is affected by a variety of factors, including weather, barriers, and user mobility, making it difficult to anticipate and optimise performance consistently.
d. **Hardware Limitations:** Physical limits in hardware, like as antenna capabilities and computing power, can limit the possibility for capacity and coverage expansion.
e. **Scalability and Complexity:** With the deployment of more base stations and innovative technologies, networks evolve and become more complicated, making it difficult to manage the increased number of parameters and related computing overhead.
f. **User Distribution and Traffic Patterns:** Uneven user distribution and dynamic traffic patterns can create hotspots and areas with low coverage, necessitating adaptive solutions.
g. **Cost of Deployment:** Increasing cell density to boost capacity and coverage can be costly, necessitating careful analysis of return on investment.
h. **SINR Handling:** Signal-to-Interference-plus-The Noise Ratio (SINR) is an important indicator for network performance, and its fluctuation owing to diverse types of clutter and propagation models affects coverage design.

When analysing the performance and capacity of heterogeneous networks, a unique set of challenges and problems emerge. For typical homogeneous cellular networks, the cellular industry has developed an elaborate approach for system capacity modelling and evaluation. The primary focus has been on common network coverage (e.g., 95%), overall cell throughput and spectral efficiency, and tail user throughput (e.g., 95% point on the CDF curve). A CDF (Cumulative Distribution Function) curve in the context of LTE (Long Term Evolution) networks is a graphical representation that displays the likelihood that a specific value (e.g., throughput, signal intensity, delay) will be less than or equal to a given number on the x-axis. It gives information about the distribution and features of network performance measures.

However, the IT sector has its own approach to network modelling and evaluation, with a focus on peak data rate and hotspot traffic density. The cellular ecosystem has broadened its traditional paradigm to incorporate femto/small cell deployment scenarios for heterogeneous networks. This includes the proportion of overall throughput carried by low-power cells, the median user throughput, and the average macrocell area spectral efficiency. Furthermore, latency-based indicators were used to evaluate bursty traffic. Novel wireless connectivity and use cases added to the cellular paradigm include closed subscriber group (CSG) and relay, as well as in-band self-backhauling [15]. Nevertheless, determining the performance and capacity of heterogeneous networks still necessitates substantial study and development. Many important areas of focus include: a unified or hybrid cellular and Wi-Fi evaluation methodology that allows for more TCO modelling; the impact of network coupling and interworking performance on subscriber distribution across network layers; the inclusion of new connectivity types such as mobile relay and device-to-device communications; cross-network-layer interference coordination and radio resource management; and more precise modelling of the unlicensed band radio.

1.6 Future Heterogeneous Network Applications

Heterogeneous networks were initially used for traffic offloading and capacity management. However, innovation within the industry has brought forth unparallelled user experiences while also providing a real computing continuum. Traffic offloading technologies are expected to evolve and improve to alleviate congestion in both access and core networks. This is accomplished by incorporating local breakout solutions like as LIPA (Local IP Access) and SIPTO (Selected IP Traffic Offload), in addition to IP flow mobility .

As backhaul transport costs rise, we expect to see an increase in the number of new and current hybrid systems that use mesh networks, in-band relays, copper cables, microwaves, optical fiber, and multihope relays. As user devices traverse increasingly across network tiers and coverage zones of different mobile network providers, network security and user privacy will need to improve greatly. Additionally, we believe that the wireless industry will devote significant research and development resources to addressing the design and deployment issues of extremely dense heterogeneous networks in public venues such as stadiums and mass transportation hubs. Collaborative communication will be critical in the development of heterogeneous networks to enable new kinds of connectivity and associated new economic models. These include sensor hubs for machine-type communication devices, client and mobile relays, and device-to-device communication. Because cognitive radio may give an additional connection option for group communication, it might be incorporated in this architecture.

Future heterogeneous networks are expected to completely incorporate and leverage next-generation air interface features. The employment of carrier aggregation across diverse network is a notable example. Without carrier aggregation, a mobile device may have to switch between macro and small networks often in the deployment scenario represented in Figure 1. 4. This causes service disruptions and an increase in network signal traffic. Carrier aggregation enables secondary data channels from small cells to operate at a higher frequency band to provide high data rates and enhanced network capacity in an opportunistic manner, while the primary control channel can be anchored at the macrocell and ideally operate at a lower frequency band for reliability and coverage.

If the mobile device is inside the macrocell's service region, no handover is necessary. This is displayed on Figure 1.5. Unrestricted access and closed subscriber group (CSG) cells, as well as interference control between various power class cells, are also made possible by several carriers. While dynamic radio resource management and network load balancing can be implemented by sharing those carriers among cells through cross-network-layer coordinated multi-point transmission and beamforming, application scheduling, power control, inter-cell and inter-network-layer interference coordination and alignment, such as fractional frequency reuse (FFR) and time-domain resource partitioning, long-term resource partitioning can be accomplished by exclusively allocating carriers to specific power class cells. Without carrier aggregation, a user device may need to handover frequently between two layers of networks, disrupting services and adding significant amount of signaling traffic.

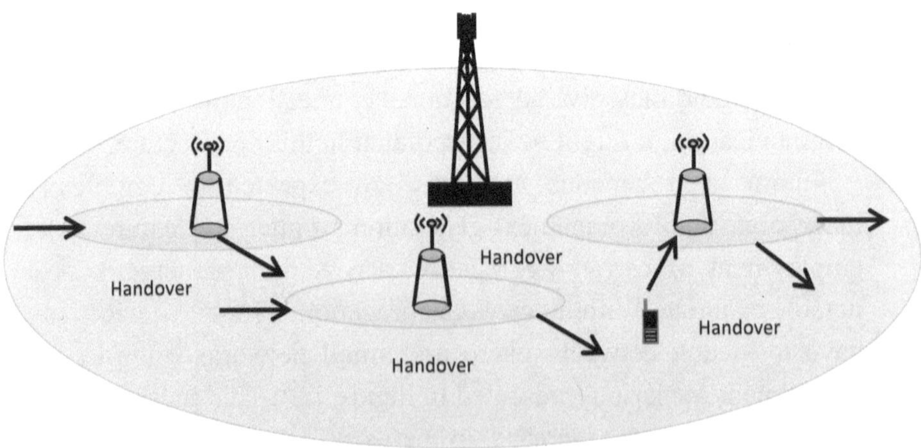

Figure 1.4 No Carrier aggregation in a heterogeneous network.

With carrier aggregation, the primary control channel can be anchored at the macro-cell, while the secondary data traffic channels operate from the small cells for high data rate and system capacity.

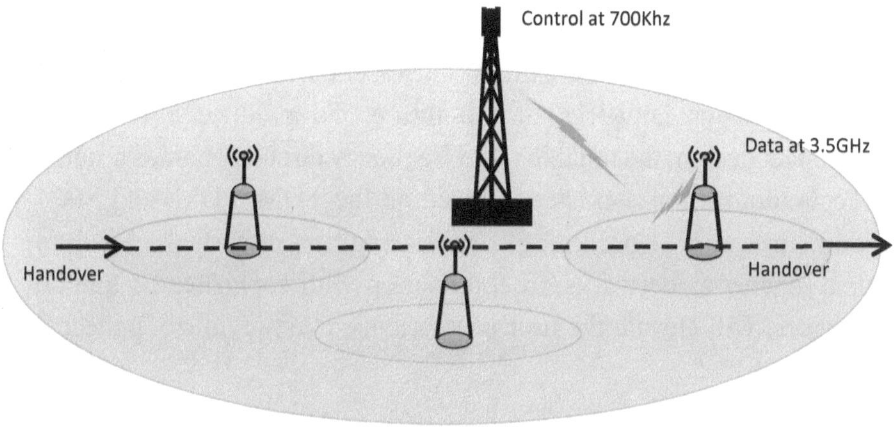

Figure 1.5 Carrier aggregation in a heterogeneous network.

Media optimization using information and communication technology (ICT) is expected to result in a considerable improvement in network performance as wireless technology shifts from spectrum efficiency to network efficiency. Providing uploads and dissemination of social networking content, local navigation mapping information, and video content. Virtualization technology enables a single signal processing platform for both the control and data planes. It also enables dynamic balancing between application processing and signalling. Future heterogeneous networks may equally benefit from the introduction of ICT in a variety of ways, including the coexistence of small cells and super-sized cloud base stations, the potential introduction of a super control node that coordinates the operation of various network layers, the expansion of data plane signal processing for cross-network QoS support, and further flattening of the core network, which would make cross-layer optimization easier to implement .

It is no longer unusual to encounter a user device with seven or more antennas as the world becomes more wireless and linked. The devices themselves constitute an essential component of the heterogeneous networks of the future. To produce a low-cost, compact design, the tunable radio frequency circuit becomes crucial. Technologies for coexistence based on the TDM, FDM, and SDM domains are becoming increasingly important. Future heterogeneous networks, in conjunction with sophisticated mobile devices, will elevate the user's experience to a previously unheard-of degree.

References

1. Geng Wu, Shilpa Talwar, Kerstin Johnsson, Nageen Himayat and Kevin D. Johnson, 'M2M: From Mobile to Embedded Internet', IEEE Communications Magazine, April 2011.
2. Shu-Ping Yeh, Shilpa Talwar, Geng Wu, Nageen Himayat and Kerstin Johnsson, 'Capacity and Coverage Enhancement in Heterogeneous Networks,' IEEE Wireless Communications, June 2011.
3. China Mobile, 'C-RAN, The Road Towards Green RAN,' Version 2.5, October 2011.
4. A. Ahmed, Z. El Saraf, F. A. Khan, and Q. Z. Ahmed, "Cooperative non-orthogonal multiple access for beyond 5gnetworks," *IEEE Open Journal of the Communications Society*, vol. 2, pp. 990-999, 2021, doi:10.1109/OJCOMS.2021.3075081.
5. ABI, Mobile Operator CAPEX Market Data, 27/07/2025.
6. K. Johansson, A. Furuskar and C. Bergljung, 'A Methodology for Estimating Cost and Performance of Hetero-¨ geneous Wireless Access Networks,' PIMRC '07.

7. K. Johansson, J. Zander, and A. Furuskar, 'Modelling the cost of heterogeneous wireless access networks,' ¨ *Int. J. Mobile Network Design and Innovation*, Vol. 2, No. 1, pp. 58–66.
8. Caroline Chan and Geng Wu, 'Pivotal Role of Heterogeneous Networks in 4G Deployment', ZTE Technologies, Issue 1, 2010.
9. Aleksandar Damnjanovic et al., 'A Survey On 3GPP Heterogeneous Networks', IEEE Wireless Communications, June 2011.
10. Klaus Doppler, Cassio B. Ribeiro and Jarkko Kneckt, 'Advances in D2D Communications: Energy Efficient′ Service and Device Discovery Radio', 2011 2nd International Conference on Wireless Communication, Vehicular Technology, Information Theory and Aerospace & Electronic Systems Technology (Wireless VITAE), 28 February to 3 March, Chennai, India.
11. Seppo Ham¨ al¨ ainen, Henning Sanneck and Cinzia Sartori, ¨ *LTE Self-Organizing Networks (SON): Network Management Automation for Operational Efficiency*, John Wiley & Sons, ISBN 1119970679, 7 February 2012.
12. GSMA, WBA Collaborate on Wi-Fi Roaming,' http://wirelessweek.com/News/2012/03/gsma-wba-collaborateon-wifi-roaming/.
13. Rose Qingyang Hu, David Paranchych, Mo-Han Fong and Geng Wu, 'On the Evolution of Handoff Management and Network Architecture in WiMAX,' IEEE Mobile WiMAX Symposium, Orlando, USA, pp. 144–149, March 2007.
14. Miikka Poikselka, Harri Holma, Jukka Hongisto and Juha Kallio, ¨ *Voice over LTE (VoLTE)*, John Wiley & Sons, ISBN 1119951682, March 13, 2012.
15. 3GPP TR 36.814, 'Evolved Universal Terrestrial Radio Access (E-UTRA); Further advancements for E-UTRA physical layer aspects. ("3GPP TR 36.814 – Evolved Universal Terrestrial Radio Access (E-UTRA").

2

Capacity and Coverage Enhancement

Network providers must make disruptive advancements in mobile broadband system architecture to meet the exponential growth in mobile traffic demand, which is now yielding flat revenues per bit. The introduction of sophisticated client devices (such as smartphones, laptop computers, and tablets) to the market has resulted in an exponential increase in mobile traffic, increasing traffic load and restricting the capacity of current wireless networks. Network providers must contend with this demand while maintaining a consistent income per bit. To meet the anticipated capacity demand at a reasonable cost, major improvements to mobile broadband system architecture will be required. One of the most promising concepts for increasing area capacity and improving indoor coverage at a low cost is heterogeneous multi-tier network architecture.

The heterogeneous network architecture described here consists of several types of infrastructure pieces, one or more radio access protocols, and a hierarchy of tiers (also known as footprints). These networks components collaborate to address a variety of difficulties related to capacity, coverage, and service quality. Because they offer the greatest possible capacity at the lowest cost, the single-frequency multi-tier deployment and multi-radio multi-tier deployment scenarios are crucial. As a result, we focus on these applications, explaining the challenges and recommending some promising interference reduction strategies. This chapter also discusses future research directions in this broad field of study, as well as industry advancements and standards.

2.1 Introduction

As more people utilize rich multimedia devices and services, the need for wireless data capacity is expanding dramatically. According to recent studies, the rise in wireless traffic has been an order of magnitude greater than the improvements in spectral efficiency (or capacity) required to meet demand. The rise in usage will continue as more people acquire smartphones and newer devices enable them to access more multimedia content. Indeed, networked computing devices like laptops and tablets can require up to 450 times the capacity of a basic cell phone, while smartphones are estimated to use 30 times that amount. The increased popularity of laptops and netbooks will add to the megatrend of mobile broadband data growth.

Despite increased traffic, operators' revenues per bit are flattening due to the prevalence of flat-rate data-centric contracts. Operators must therefore figure out how to increase capacity at a far lower cost per bit. When employing typical network design methodologies, mobile service providers have three primary tools for increasing capacity: cell splitting, improved air interface, and newer spectrum. Because each cell split may necessitate additional sites, civil construction, and regulatory approval, increasing the density of the macrocell site can be costly.

Nonetheless, the spectral efficiency of 4G air interfaces has transformed to the point where it approaches system capacity limits. New spectrum purchase can be expensive, subject to government regulatory timelines, and limited in availability. As a result, the wireless industry is shifting its focus from maximizing spectral efficiency to improving the effectiveness of networks. In this regard, developing heterogeneous networks (HetNet's) offer a potential paradigm for increasing system capacity and coverage at a fair cost. Most current research in heterogeneous networks has focused on multi-tier techniques that attempt to offload as much traffic to smaller cells as possible. Although this technique has shown promising outcomes, it is only the beginning.

We regard heterogeneous networks as a platform and enabler for disruptive technical advancements, as well as critical to the advancement of mobile wireless broadband. The fundamental advantage of heterogeneous networks is their capacity to adapt infrastructure components (in terms of cost and power/range) to the environment. For example, the client the network's least expensive and most power-hungry device, can collaborate with other clients to improve local access and throughput. However, pricey, strong macro base stations (MBSs) provide seamless mobility and blanket coverage across large geographic areas.

Between these two extremes, femto access points (FAPs) lower cell radius to less than 100 meters and provide inside coverage, whilst pico base stations (PBSs) and relay stations (RSs) boost coverage and capacity with smaller cells of several hundred meters in radius. Low-power network nodes (FAPs, PBSs, RSs, and clients) can significantly boost coverage and area capacity by shortening transmission distances and improving spectrum reuse. Because heterogeneous installations are multi-tiered, they are ideal for using spectrum across many radio access technologies, including WLAN, to boost capacity.

Operators can offload enormous amounts of data to unlicensed bands by including WLAN (such as Wi-Fi) into a cellular network [1, 2]. Consumer gadgets and infrastructure are rapidly integrating multiple radio technologies to make use of this alternative spectrum. Heterogeneous networks provide cost benefits in addition to higher capacity. In terms of strategy, the provider can save a significant amount of revenue by adopting low-cost network components to meet coverage and capacity requirements rather than purchasing more spectrum or deploying denser macro/microcells. These components are not only less expensive, but site acquisition and backhaul costs are sometimes reduced or eliminated.

In addition to that, using free, unlicensed spectrum judiciously reduces costs even further. This chapter summarizes the architecture of heterogeneous networks and the accompanying protocol solutions. We then discuss the most difficult but most cost-effective method of increasing capacity with multi-tier deployments, which is to reuse the same carrier throughout the network's multiple tiers to make the best use of limited spectrum. We show that the purported multi-tier capacity and coverage benefits of this deployment can only be achieved with effective interference management solutions that limit cross-tier interference.

2.2 Small Cell Access Mechanism

Femtocells are minuscule cells, and there are three primary methods for accessing them.

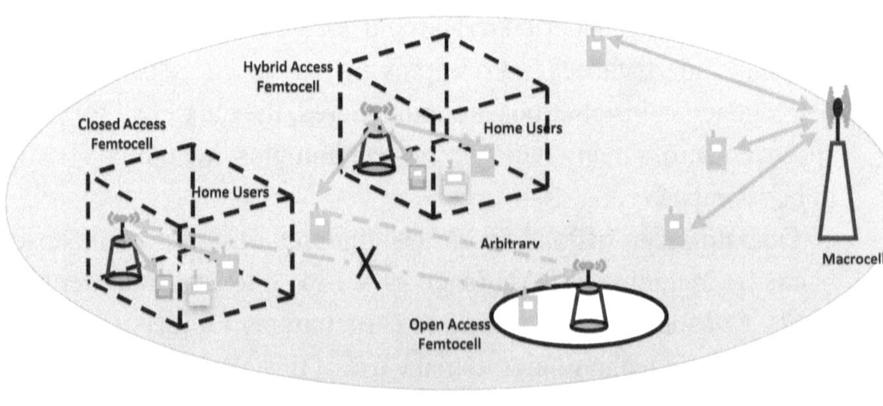

Figure 2.1 Approaches to access a small cell (i.e., femtocell).

A femtocell with limited access is only accessible to a predetermined number of registered users, most of whom are at home (Figure 2.1). In an open access scenario, any number of cellular operator customers who live closer to a femtocell but far from a macrocell can connect to it. However, a deployment technique known as hybrid access is employed when some resources are assured for other users but restricted for registered home users.

Closed access is therefore excellent for home users who want to retain network privacy while still ensuring that there is sufficient capacity to manage traffic. Similarly, open access benefits the network owner by expanding its reach. Operators, however, favour hybrid access to prevent lack of access when using open access to resources and to lessen the disruption that closed access causes in the uplink. subsequently the user's distance from the macrocell determines whether to set up a femtocell with open, closed, or mixed access.

2.3 In-building Signal Propagation Characteristics

2.3.1 Path Loss and Multi-path Fading

Line-of-sight components between a UE and a BS are likely to exist in interior environments. Because the channel structure does not change significantly over short distances, indoor channels in the same area are similar. If all the apartments in a multi-story building have the same structure, it is reasonable to assume that all indoor channels will have a similar shadowing effect. Doppler spread and delay spread effects on the channel can contribute to small-scale channel fading. In wireless communication, doppler spread is the term used to describe the change in frequency shifts of a received signal brought on by the relative motion of the transmitter and receiver. The received signal is dispersed over a range of frequencies instead of just one due to this phenomenon, which is called the Doppler effect. Understanding and creating dependable wireless communication systems require an understanding of doppler spread, particularly in dynamic or mobile environments.

Nevertheless, an indoor channel experiences less Doppler spread than an outdoor channel, leading to a long channel coherence time, partially due to the relatively limited motion of UEs and objects between a UE and BS. Additionally, the delay spread of an indoor channel is usually less than 100 ns. Indoor channels are less susceptible to small-scale fading effects and are less likely to change significantly inside a given space because they have smaller delay and Doppler spreads. On both large and small scales, indoor channels within a structure usually undergo comparable fading effects.

Line-of-sight components between a UE and a BS are likely to appear in interior settings. Because the channel structure does not change much over short distances, indoor channels in the same area are similar. If all the apartments in a multi-story building have the same structure, it is logical to expect that all indoor channels will have a similar shadowing effect. Furthermore, Doppler spread, and delay spread effects on the channel result in small-scale channel fading.

An internal channel between a UE and BS, on the other hand, exhibits less Doppler spread than an outdoor channel due to the comparably limited motion of UEs and objects between the two. This results in a significant channel coherence time. Furthermore, a route that is indoors has a shorter delay spread, usually less than 100 ns. Because interior channels have lower Doppler spreads and delays, their properties are less susceptible to small-scale fading and do not change significantly over time.

2.3.2 Penetration Loss in Millimetre-Wave Bands

Indoor small cells are a tried-and-true solution for increasing indoor voice and data coverage and capacity. Little cells are vital in networks today and in the future. It can be simple, inexpensive, and perform well in interior settings to deliver the necessary coverage and capacity outside of the macro network.

Floor penetration loss is frequency-dependent, increasing rapidly as frequency increases, again it is not consistent across all floors between transmitters and receivers. Instead, the influence of floor penetration loss is nonlinear, decreasing as the number of floors increases. There is currently no standard methodology for measuring the penetration loss of walls and floors using various building materials [3]. Using and fitting the curve for a concrete block in, according to [4], the first-floor penetration loss in the 28 GHz mmWave band is 55 dB.

Furthermore, penetration losses through walls, both internal and external, are significant. The radio frequency is contained within a building due to the high penetration loss (at high frequencies, such as 28 GHz) for both intra-floor (caused by internal walls) and inter-floor levels, as well as the high penetration loss of external walls. This causes little or no interference with signals of the same frequency outside the building. In addition, when an outdoor signal penetrates a building's exterior wall, its strength drops, resulting in the reuse of an outdoor base station's frequency for indoor users served by indoor base stations located within the structure [5]. When an outdoor base station's frequency is reused for inside users in the same building, it causes severe co-channel interference. This necessitates proper interference control. Note that penetration losses rise in direct proportion to frequency.

To meet the high-capacity requirements of next-generation mobile systems, including 5G and beyond systems in dense urban multistorey buildings, the mmWave spectrum bands are thought to be excellent candidate bands for spectrum extension. In addition, MNOs keep lowering cell coverage to take advantage of spatial reuse. In this sense, short-range (usually 100-200 m in radius) mmWave technologies are seen as promising, especially in urban interior environments, to address the huge deployments of small cells to give high data rates at a short distance. Additionally, given the widespread availability of the mmWave spectrum, a single MNO may be unable to fully utilize it. In these cases, dynamically sharing the mmWave spectrum across multiple operators within a country can help to improve mmWave spectrum use.

2.4 Enabling Technologies

A heterogeneous network holds various radio access protocols, access tiers (also known as footprints), and related infrastructure components. The received signal capacity at a receiver is proportional to its distance from the transmitter and the available spectrum bandwidth. The received signal capacity improves as the distance decreases, and the spectrum bandwidth increases. Reduce the cell size to get the transmitter and receiver as near together as possible.

As depicted in Figure 2.2, a macrocell with radius R running at spectral bandwidth B generates little cells with radius r each. When the macrocell coverage is divided into multiple smaller ones, it is clear that the same spectrum (B where B=b) can be reused spatially (an indirect effect on spectrum extension), resulting in an increase in capacity over a specific area (i.e., CM = α *CS, where CM and CS represent macrocell capacity and total small cell capacity, respectively).It is clear that the decrease of the macrocell coverage into several smaller ones permits the spatial reuse of the same spectrum (B where B=b) if the Signal-to-Interference-plus-Noise-Ratio (SINR) is the same for the macrocell and small cells (an indirect impact toward the spectrum extension). As a result, increased capacity is achieved across a specific area (i.e., CM = α x C S, where α is the spectrum reuse factor, which is 7 in Fig. 2.3 and CM and CS stand for the macrocell capacity and total small cell capacity, respectively). From Shannon's capacity formula, the network capacity can be enhanced mainly by addressing three directions, spectrum accessibility, spectral efficiency improvement, and network densification.

$$C_L = \Phi \times B \operatorname{Log}_2\left(1+\left(\frac{P_r}{N+I_T}\right)\right)$$

where C_L, ϕ, B, P_r, N, and I_T denote, achievable capacity, spectrum reuse factor, available spectrum bandwidth, received desired signal power, noise, and received total interference signal power, respectively.

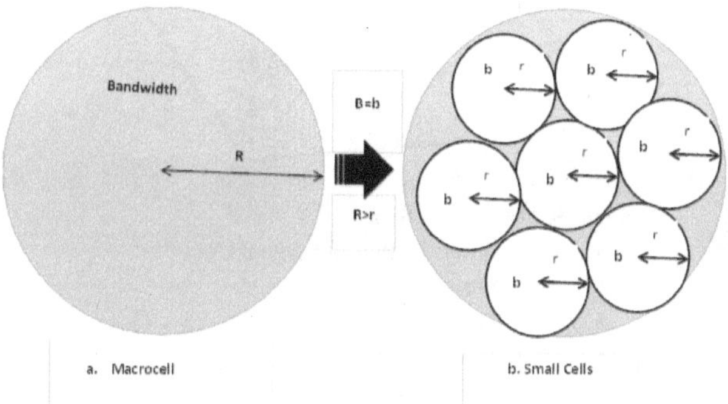

Figure 2.2 Formation of small cells from a large macrocell.

(a) A macrocell, (b) A set of small cells.

These are depicted in Figure 2.3 along with three directions in a triangle for network capacity increase. Corresponding enabling technologies are also displayed in each direction.

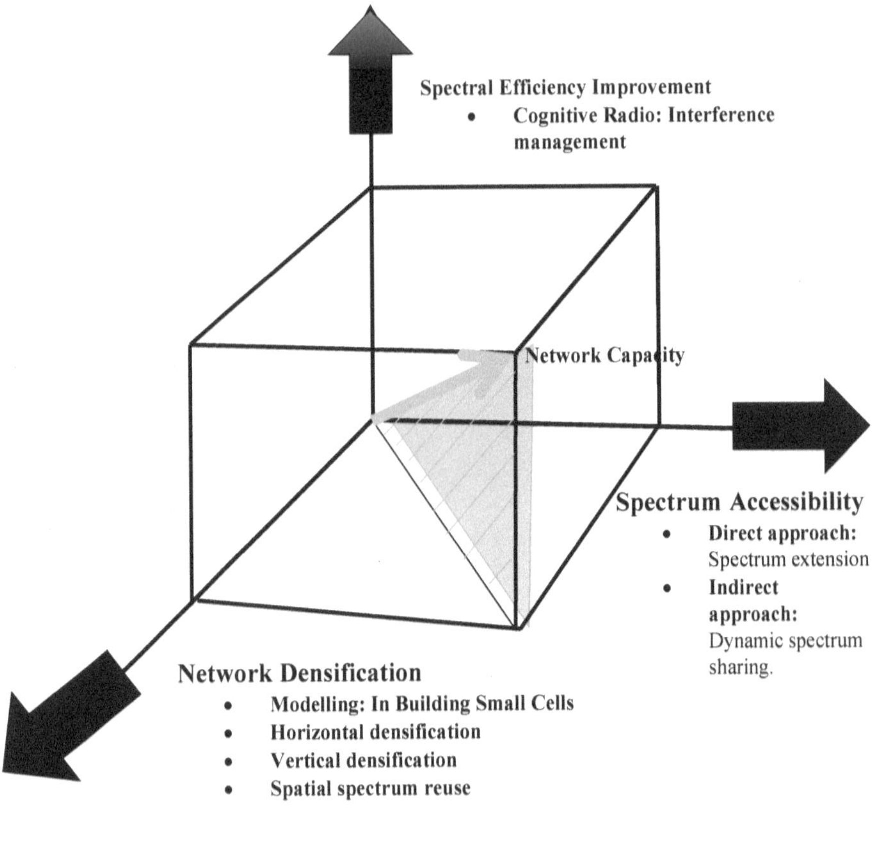

Figure 2.3 Network capacity improvement triangle.

2.4.1 In-Building Small Cell Network Densification

The process of strategically placing small cell base stations inside buildings to increase network capacity and coverage particularly in regions with high data traffic or weak signal strength is known as "in-building small cell network densification." This entails supplementing current macrocell networks with low-power cellular radios placed in places like offices, public areas, and hallways. Because of their restricted coverage and low transmission power, SBSs are regarded to be an effective means to send a significant amount of indoor data at a high data rate. Furthermore, SBSs can be installed on a building's intra-floor and inter-floor levels, resulting in an ultra-dense SBS deployment across a specific 2D physical space region within a macrocell's coverage. To capture 3D impacts on indoor signal propagation, such as floor penetration loss, it is necessary to consider a building's height.

2.4.2 3D Spatial Spectrum Reuse Indoors

Reusing the same radio frequency spectrum in various spatial locations within a building to boost network capacity and spectral efficiency is known as 3D spatial spectrum reuse in indoor environments. This is conducted by deploying small cells (SCs) in three dimensions, both vertically (inter-floor) and horizontally (intra-floor), enabling the reuse of the same frequency in various small cell clusters. mmWave frequencies experience higher penetration loss through walls (both internal and external) compared to lower frequency bands like those used by 4G or lower 5G bands. This means mmWave signals have a harder time traveling through building materials, leading to weaker signals indoors when using an outdoor mmWave source or vice versa.Because mmWave bands have a higher penetration loss through internal and external walls and floors in any multistory building than low-frequency microwave bands, they can be reused in the third dimension, or the height of a multistory building. This results in the reuse of the same mmWave band more than once at the inter-floor level. Furthermore, widespread reuse of mmWave spectra in ultra-dense deployed tiny cells within a multistory building can be eased by the standard intra-floor spectrum reuse techniques.

It should be noted that a channel's capacity is directly proportionate to its available spectrum bandwidth. This bandwidth can be enhanced by increasing the number of available spectra, such that each small cell can use many spectrums, as well as the number of times small cells in a multistory building reuse the same spectrum using vertical spatial reuse. Thus, technologies for 3D spatial reuse of high-frequency millimeter-wave spectra using in-building multiband-enabled small cells can achieve extraordinarily high capacity and data rates in accordance with user expectations for future mobile networks.

2.4.3 Spectrum Extension in Millimeter-Wave Bands

Due to the limited availability of lower frequency bands, 5G and future wireless technologies depend on spectrum extension in millimeter-wave (mmWave) bands. Although mmWave, which typically operates between 30 and 300 GHz, provides a large bandwidth for high data rates, it has drawbacks such as atmospheric attenuation from rain, fog, and even air, which can scatter or absorb mmWave signals and degrade them and a shorter-range limitation. mmWave frequencies are perfect for 5G and beyond because they enable much faster data transmission rates than lower frequencies. With the potential for even faster data rates and new uses, MmWave is an essential part of 6G and beyond.

In dense urban multistory structures, the mmWave spectrum bands have been excellent candidate bands for spectrum extension to meet the high-capacity demand of next-generation mobile systems, including 5G and beyond technologies. Additionally, to take advantage of spatial reuse, MNOs keep lowering cell coverage. The short-range (usually 100-200 m in radius) mmWave technologies are thought to be promising in this regard, especially in urban interior contexts, to address the vast deployments of small cells and to give high data rates at a short distance. Additionally, a single MNO might not be able to completely utilize the mmWave spectrum due to its immense availability. In these situations, allowing several operators in a nation to share the mmWave spectrum dynamically can help to increase its use even further.

2.4.4 Dynamic Spectrum Sharing

A more effective use of spectrum resources is made possible by 5G technology called Dynamic Spectrum Sharing (DSS), which allows 4G LTE and 5G NR to share the same frequency bands. Without requiring dedicated spectrum for each technology, it enables a more seamless transition to 5G by dynamically allocating resources between the two based on real-time traffic demands. To service the traffic of its consumers, each MNO in a nation is often given a specific licensed spectrum. In the past, these fixed distributions of radio frequencies were adequate to meet user needs. To serve more customers with realistic data rate demands, MNOs may need to utilize techniques like fractional frequency reuse to reuse the same designated spectrum. The demand for mobile communications has increased dramatically over the past ten years because of rising subscriber numbers, traffic volumes, and data rates per user keep rising. Since most data is created indoors, it has become imperative to provide high indoor data speeds and capacity per user.

The radio spectrum allotted to MNOs has not increased proportionately, even though their subscriber base has grown significantly. It has recently been established that dynamic sharing of radio spectrum that is already statically allocated to MNOs using indoor small cells is more effective. A secondary system may dynamically or opportunistically share the spectrum allotted statically to a primary system in the DSS, provided that the primary system is not impacted by the sharing. Utilizing their structures, indoor small cells can be essential to the realization of several DSS methods. In this sense, the Almost Blank Subframe (ABS)-based eICIC can be used in small cells to provide time orthogonality while serving the traffic of the corresponding MNO to prevent CCI when sharing the licensed spectrum of two MNOs.

2.4.5 Cognitive Radio

Cognitive radio is a type of radio communication that can intelligently assess its surroundings, including frequency spectrum usage, and adjust transmission parameters to improve performance while minimising interference. It is a type of dynamic spectrum management that enables the more efficient and effective use of available radio frequencies. This makes it possible for a nation's MNO to effectively use its licensed spectrum is another significant difficulty. This is because the demand for the necessary quantity of spectrum for various MNOs varies in tandem with the sudden changes in the user traffic demand of many MNOs in a nation throughout time and location. This results in a significant amount of spectrum that is allotted to each MNO in a nation being underutilized or left unused in both time and space.

Considering that most data is created indoors, especially in crowded, multistory metropolitan buildings, providing huge amount of data indoors makes the difficulties much more pressing. For an MNO to maximize meeting its increased user demand at high data speeds, especially in congested metropolitan multistory buildings, it is imperative that it both increase its available spectrum and improve its utilization of that spectrum. To address this issue of underutilization of the spectrum, Cognitive radio (CR) has emerged as an enabling technology in recent years.

One of the main functions of Cognitive radio (CR) is spectrum access, which keeps Primary UEs (PUs) and Secondary UEs (SUs) from colliding and permits sharing of an MNO's licensed spectrum with another to boost its effective spectrum bandwidth and support high data rates and capacity. The following sections go into detail about each of the three paths in the network capacity improvement triangle depicted in Figure 2.4 (i.e., network densification, spectrum accessibility, and spectrum efficiency improvement) by using the corresponding enabling technologies previously mentioned to increase the in-building capacity.

2.5 Base Stations

LTE base stations, also known as eNodeBs in LTE (Long Term Evolution) networks, serve as the primary communication hubs that connect mobile devices to the core network. They enable wireless communication between mobile devices and the larger network. Previously, most base stations were wide area macro cells, which used high-mounted antennas and high-power transmitters to span many square kilometres. Outdoor small cells, also known as micro cells, cover a few hundred meters with lower antenna placements. Indoor cells can be found in homes or in public areas such as office buildings or shopping malls. The public indoor solutions, known as Distributed Antenna Systems (DAS), have typically used internal antennae and RF lines coupled to a macro base station located within the same structure.

A DAS has the advantage of making it relatively simple to upgrade new radios; in the best-case scenario, all that is required is inserting a new base station into the cable system without modifying the cables or antennas. The disadvantage of DAS is that, because of a larger coverage area it simply shares the same physical layer transmission resources, adding more antennas may only improve coverage and provide a more uniform quality of service throughout the coverage area; it will not result in significantly higher capacity. The only method to increase capacity is to separate cells, which is a complex procedure that will certainly necessitate extensive cabling changes.

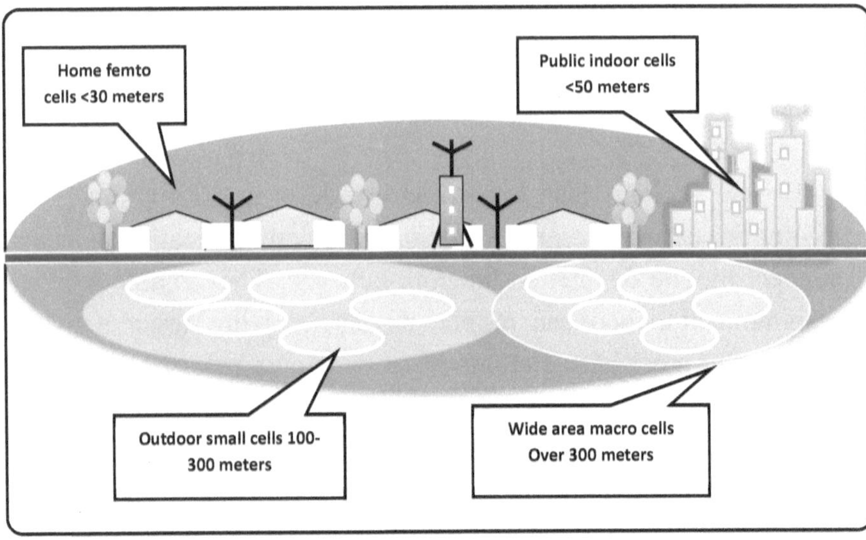

Figure 2.4 Outline of the Various Cell Types

Another indoor solution is a home base station, also known as a femto cell. The fundamental goal of femto cells has been to boost capacity and coverage in homes and small offices. Figure 2.5 depicts an outline of the various cell types. To accommodate users with diverse needs, future mobile networks may stack devices with varied footprints and capabilities within the same geographic area. Devices on the same network utilize the same radio access protocols and operate at the same frequency. The multi-tier design allows for substantial spectrum reuse, which improves connection quality and capacity. Here, we discuss the function of each infrastructure item as well as the tier associated with it. We also provide examples of key usage scenarios.

2.5.1 Macrocells/microcells

The 3GPP has defined different RF requirements for home base stations, local areas, medium range networks, and wide area networks. The ability of users to use antennas provisions is important in base station classification. This is measured as MCL, or Minimal Coupling Loss. For Wide area base stations, the MCL is more than 70 dB, which corresponds to a high-mounted antenna installation on a rooftop, tower, or pole. Local area base stations, which let users to go remarkably close to the antenna, are characterized as having an MCL or Minimal Coupling Loss of 45 dB, while medium range base stations assume an MCL or Minimal Coupling Loss of more than 53 dB. The maximum power level of a wide area base station is not limited in 3GPP specifications.

The conventional 2 x 2 MIMO solutions support up to two 80 W power levels. Furthermore, macro base stations with three to six sectors often use sectorization to enhance site power to 1 kW. Medium-range, local-area, and home base stations are limited to 5 W, 0.25 W, and 0.10 W, respectively. These power levels show the aggregate power of all the transmission levels. For example, in the case of two or four antennas, the 0.25 W limit is equivalent to 2 x 0.125 W or 4 x 0.0625. Other variables in RF needs include sensitivity, dynamic range, spurious emissions, frequency stability, and blocking. The cost of RF sections can be further reduced because small base stations often have less strict RF requirements than high power base stations. Table 1.1 shows the distinct kinds of LTE base stations.

In a mobile network, there are two types of cell coverage: macrocells, which provide a larger coverage area, and microcells, which concentrate on denser or more specialized locations. While microcells are smaller and often used in cities, shopping malls, and transit hubs, macrocells are frequently large and serve areas like roadways or rural areas. Demands for traffic have an impact on the MBS footprint. Microcells with smaller cell radii are required in urban areas, while macrocells with site-to-site distances greater than 500 m are required in rural or suburban areas. In a variety of networks, micro and macrocells offer crucial wide-area coverage.

They can also help users who have a lot of movement, which lessens the need for frequent handovers.

Table 2.1 3GPP base station classes

3GPP Base Station	Minimal Coupling Loss	Use	Deployment	Backhaul
Macro BTS	• MCL > 70 dB • Typical up to 100 W per sector (no upper limit), 3-6 sectors	Wide area base station	Operator Deployed	Wired Backhaul or Dedicated Wired
Micro BTS	• MCL > 53 dB • Max 5 W	Medium range base station	Operator or Community Deployed	Wired Backhaul or Dedicated Wired
Pico BTS	• MCL > 45 Db • Max 0.25 W	Local area base station	Enterprise or operator deployed	Shared or dedicated backhaul
Femto BTS	• Max 0.10 W	Home base station	User Deployed	Shared backhaul

2.5.2 Picocells

These cells have an inter-site spacing of between 100 and 300 meters. Picos are used in places with limited capacity, such stadiums, airports, and retail centers, Pico base stations (PBSs) can be set up as hotspots. PBSs are less expensive and more power-intensive MBSs. Since they are usually a part of the operator's public infrastructure, all network users can access them. Operators typically carefully arrange the deployment of PBSs.

2.5.3 Relays

LTE Relays, or relay nodes in LTE-Advanced (LTE-A) networks, are devices that help extend coverage and increase throughput, particularly at cell boundaries, by transferring signals between a base station (eNodeB) and user equipment (UE). They are an essential element of LTE-A that improves network speed and capacity without requiring a complete base station deployment. The need to increase LTE coverage is clear. More macro or small base stations can be placed to improve coverage. The additional base stations increase both coverage and capacity. Finding suitable base station site locations and establishing efficient transmission of signals is often a hurdle for new base stations. Using RF repeaters, which enhance signal strength but also increase interference, has been another method of expanding coverage. Controlling interference with repeaters is challenging in today's networks. Already with 3GPP, relays become an additional option for improving network coverage.

The relay node's idea is to use the same LTE air interface for the backhaul connection, which simplifies relay installation by eliminating the need for a separate transport solution. Relay stations cover areas comparable in size to PBSs. Portable LTE Base Stations used in Public Safety Broadband Networks (PSBN). These portable base stations are used to expand or enhance existing network coverage, especially in emergency situations or locations with inadequate infrastructure. They are an important part of Public Safety LTE (PS-LTE) networks, allowing crucial communication for first responders. On the same note, relay stations provide macro/microcells with greater coverage and throughput by forwarding traffic between base stations and mobile stations. Landline resources are unnecessary because RSs and BSs communicate wirelessly. This limits the amount of spectrum that may be accessible, however. When wired backhaul is inaccessible or difficult to reach, RSs are often installed in regions previously served by PBSs.

2.5.4 Femtocells

Femto access points cover small locations, such as homes or apartments, ranging from 10 to 50 meters in size. FAPs save infrastructure costs by using existing household backhaul, such as DSL or cable, as opposed to MBSs or PBSs, which connect to the network using operator-owned backhaul. Furthermore, FAPs are often privately owned and can be deployed more effectively by a user based on his requirements, such as in locations with insufficient indoor coverage. When a FAP shares spectrum with other network tiers, it might cause considerable interference for neighbouring users who do not member to its access group. FAP access is typically restricted to a certain subscriber group (closed group).

2.5.5 Client relays

LTE client relays, also known as client-relay, are a technology used in LTE networks to improve data rates and coverage, particularly for mobile devices near the cell edge or in coverage gaps. In this technique, some mobile devices (relay-MNs or R-MNs) with acceptable signal quality serve as relays for other mobile devices (client-MNs or C-MNs) that have poor signal conditions. This enables higher data rates and overall network performance in places where traditional base station coverage is limited.

In the design of the heterogeneous network, client relays add another tier. This tier is the only one that is not supported by infrastructure components because it is situated between clients. Links at this class are noticeably short between 10 and 20 meters. By using the strong link between a participating client and a base station (BS) (which may be macro, micro, or pico), client cooperation (CC) increases the likelihood that an end-client with a weak link to that BS can successfully transmit. In practice, CC enhances the quality of virtual links for users in low-quality areas (i.e., cell-edge users), which leads to a notable decrease in the amount of battery power and channel resources they utilize, as well as the amount of interference they cause to neighbouring cells. Research shows that CC can increase the average network throughput by 80–200% [6].

2.5.6 Distributed antennas

A distributed antenna system (DAS) is made up of several antennas scattered across a coverage area that are connected to a central hub (the headend) via a transport medium such as fibre optic or coaxial wire. This enables more efficient signal distribution and management than depending on a single, central antenna. DAS can collect signals from a variety of sources, including off-air antennas (which capture signals from existing cell towers), small cells, and even the carrier's network. The signal is then sent over the transport medium (fibre or coax) to the remote units (or nodes), and finally to the individual antennas. These antennas are purposefully placed to ensure maximum coverage and capacity in the given area.

In addition to the diverse range of infrastructure devices and capabilities, new radio technologies add new tiers to the heterogeneous network architecture. The distributed antenna system (DAS) is a prime example of this [7]. After spatially separating the antennas of a standard base station (BS), DAS connects them to a central processing unit using a fast transmission line such as optical fiber. In effect, it transforms a collection of smaller cells into a macrocell. DAS enables the operator to replace its high-power centralized antennas with several low-power antennas that cover the same cell area. Because (near) line-of-sight channels are commonly present, DAS uses less power, which enhances link quality, coverage, and capacity.

2.6 Spectrum Accessibility

High-frequency millimetre-wave spectrum bands have already been considered to satisfy the high-capacity requirements of 4G/5G and future mobile systems, particularly indoors in multistory structures, while spectrum bands below 3 GHz are totally occupied. As a result, the short-range and abundant mmWave spectrum hold promises for large-scale deployments of small cells to provide high data rates over short distances, particularly in urban indoor situations. There are two major approaches for expanding an MNO's spectrum available: The direct approach by adding (licensing) new spectrum statically and the indirect approach: by sharing used spectrum dynamically/ opportunistically.

The direct approach allows for the direct addition of new licensed spectrum, whether contiguous or non-contiguous, to a mobile system via technologies such as carrier aggregation. However, the restricted availability of radio spectrum, particularly below 3 GHz [8], and the prohibitive cost of spectrum licensing make traditional direct techniques of spectrum expansion inefficient. To address MNOs' increasing indoor high data rate and capacity demands, indirect methods ought to be used. The indirect technique enables a secondary system to dynamically or opportunistically share spectrum that is already in use by a primary system, if the primary system is unaffected by the sharing the spectrum. This type of approach is known as DSS. Indoor small cells can play an essential role in DSS.

Several small cell base station architectures can be used to manage various DSS approaches, depending on the number of physical transceivers and the size, quantity, and characteristics of an SBS's operational spectrum.

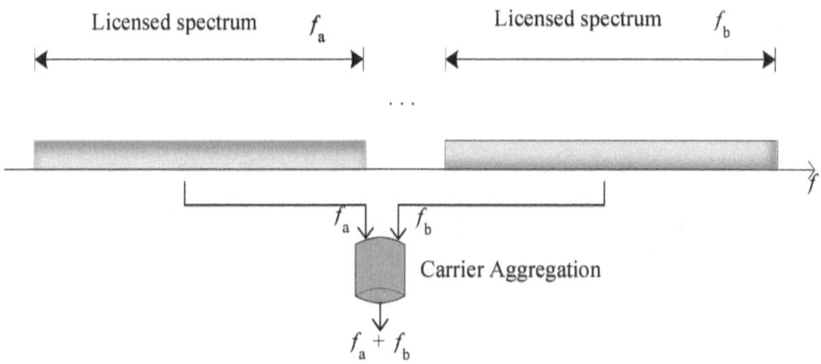

Figure 2. 5 Spectrum access using the carrier aggregation approach.

To be more particular, in [9], several DSS approaches, such as Co-Channel Shared Access (CSA), Licensed Shared Access (LSA), Unlicensed Shared Access (ULA), Authorized Shared Access (ASA), Co-primary Shared Access (CoPSA), and Licensed Assisted Access (LAA), are realized by enabling SBSs with a single-/multiple-transceiver and operating them at either a single or multiple licensed or unlicensed spectra of homogeneous or heterogeneous systems.

2.6.1 Static Interference Avoidance through Frequency Reuse Patterns

The simplest way to tackle interference issues when an operator has many carriers or frequency bands available is to deviate from a co-channel deployment and adopt specific frequency reuse patterns. The most basic of them is "hard frequency reuse," in which small and macro cells simply use distinct frequency bands, hence avoiding all the previously listed interference constellations. Few operators, however, have the luxury of having enough spectrum available for hard frequency reuse; even if they have, they may consider recycling.

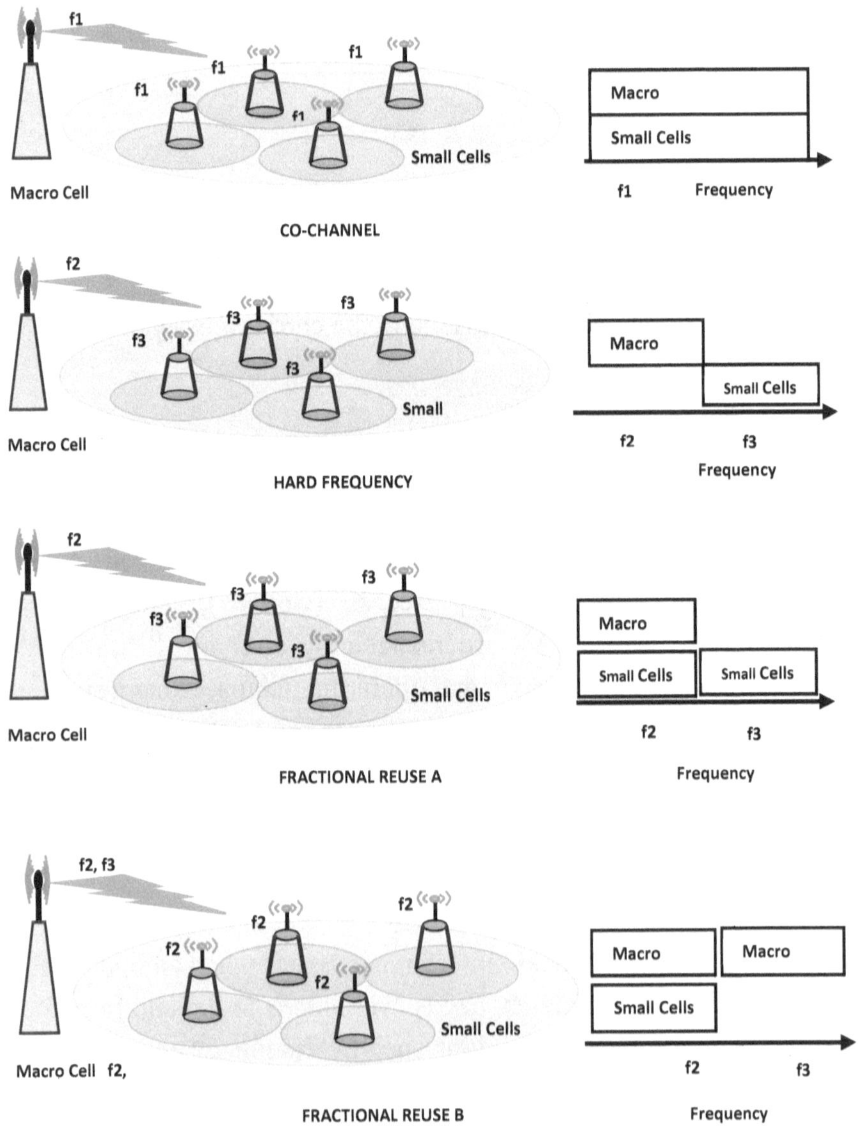

Figure 2.6 Frequency reuse options.

it all for both big and small cells to maximize capacity.

Fractional frequency reuse is an alternative technique in which some resources are allocated to a single layer while others are shared by all layers. In one variant of this layout, small cells may employ both f2 and f3, while macro cells only use f2. The location of a specific terminal connected to a small cell can then select which frequency to use for that terminal: This can be assigned to frequency f2 if it is in the core of the small cell, but if it is close to the macro-small cell border, it should only consume resources within f3 to prevent compromising macro cell performance.

A different kind of fractional frequency reuse is one in which a frequency often called an escape carrier is set aside exclusively for use by macro cells. This idea is especially intriguing for avoiding the previously described interference constellation C since UEs that are near a CSG femto cell but are unable to connect can easily be served by the macro cell and on the escape carrier, preventing interference issues [10]. Figure 2.6 shows all the frequency reuse possibilities.

2.6.2 Dynamic Interference Coordination in the Frequency Domain

Dynamic interference coordination (IC) in the frequency domain is a wireless communication approach for managing and mitigating interference by intelligently assigning and adjusting radio resources across several frequency bands. It optimises spectrum efficiency and network performance by dynamically adjusting frequency reuse patterns in response to traffic and interference levels. Rather than assigning predetermined frequency bands to cells or users, dynamic IC changes frequency allocation in real time. It adjusts to changing traffic patterns, user locations, and interference levels to ensure optimal resource utilisation. The purpose is to reduce interference between neighbouring cells and users, resulting in higher signal quality and network capacity. This includes strategies to maximize spectrum use and minimize interference while preserving network performance, such as resource partitioning, dynamic channel allocation, and frequency reuse.

Inter-cell interference coordination is the general term for the strategy (ICIC). There are two main problems with this method: On the one hand, the latency of information transferred over the X2 interface slows down base station coordination; thus, the gain of rapid frequency selective scheduling must be balanced against the interference coordination gain. However, because control channels are always delivered by all base stations irrespective of resource utilization, interference cannot be reduced by an interference-aware scheduling of data transmissions. To overcome both issues, enhanced ICIC (eICIC), a time-domain method of dynamic interference coordination, was introduced.

2.6.2 Dynamic Interference Coordination in the Time Domain

The time domain eICIC technique is a well-known example in which a BS (usually a macro BS) mutes its transmission during particular subframes known as Almost Blank Subframes (ABS) so that neighbouring BSs (typically small cells) schedule their edge users in the protected subframes. Low power nodes (LPNs) such as Femto, Pico, and Relay are deployed with overlapped coverage with macros in cochannel HetNet. The Cell Range Expansion (CRE) approach is used to offload heavy traffic from Macro cells in Macro/Pico configurations [10]. However, interference between two layers of nodes is difficult to control due to uneven transmission power; Pico UEs are particularly vulnerable to DL interference. While the correct design of virtually blank subframes (ABS) pattern is up to implementation, one of the fundamental methods established in 3GPP [3-6] is the time-domain enhanced inter-cell interference coordination (eICIC) technique.

The fundamental idea is that Almost Blank Subframes (ABS) are used to partially muffle certain subframes. An ABS is a subframe in which no data is carried, only necessary common channels and common reference signals (CRS) are sent. As a result, a cell that uses ABS transmission produces significantly less disturbance to the cells around it. The working principle between macro and pico cells is shown in Figure 2.7. ABS is allotted by a macro cell and can be utilized by the pico cell to service UEs located farther away. Other subframes can be reused by both macro and pico cells; for instance, pico cells can use them to service UEs that are closer to them.

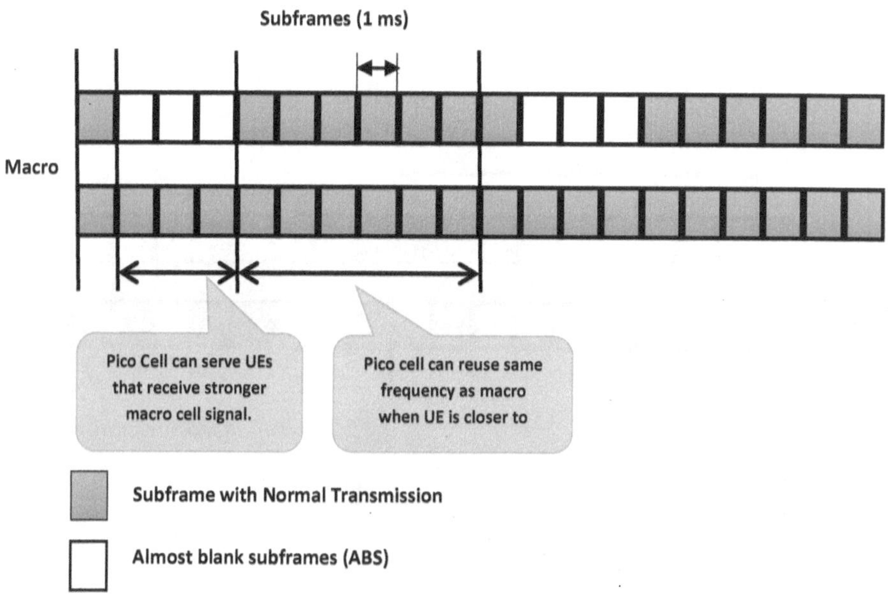

Figure 2.7 Enhanced ICIC between macro and pico cells.

The use of ABS facilitates pico cells servicing more UEs, but it also reduces the capacity of macro cells. Figure 2.8 illustrates the use of ABS between a macro cell and a CSG femto cell. In this case, femto cells employing ABS still allow macro users to receive signals when they are close by [10]. Enhanced ICIC goes as far as requiring coordination between macro and small cell coordination regarding ABS muting. The changes on ABS settings are possible through an X2 interface, which we assume exists between macros and pico cells, but not between macros and femtocells. The femtocell ABS configuration can be semi-static, controlled by the O&M center as shown in Figure 8.13. The X2 application protocol (AP) version 10 provides additional information elements that allow operators to configure ABS muting patterns between eNodeBs. While 3GPP has standardized the new X2 signaling, the eNodeB algorithms are vendor-specific and not standardized.

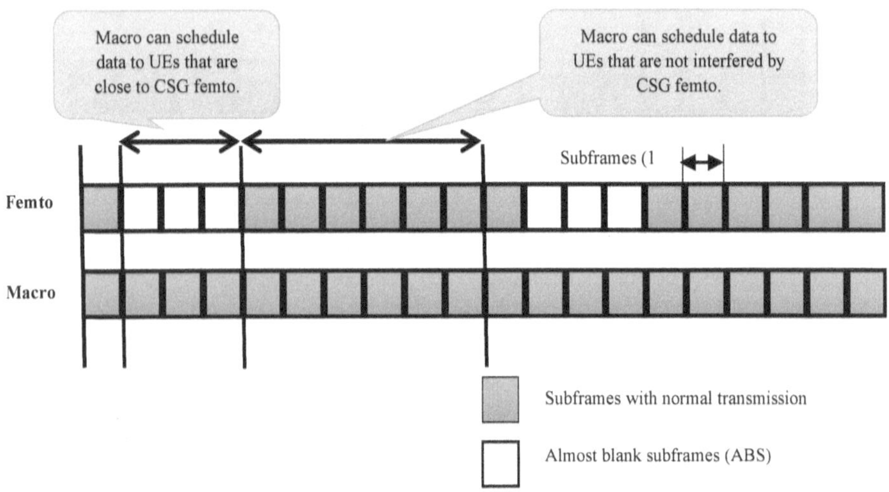

Figure 2.8 Enhanced ICIC between macro and Closed Subscriber Group (CSG) femto cells.

In contrast to rapid Radio Resource Management (RRM) within a single eNodeB, the typical adaptation of ABS muting patterns between eNodeBs is slow. ABS modifications take hundreds of milliseconds or a few seconds, but RRM actions have millisecond resolution. Over X2, the ABS enhancements produce a negligible control signaling load. The terminal must also know which frames are muted for ABS to function effectively. 3GPP Release 10 was intended to be backwards compatible with Release 8 User Equipment (UE) and introduced Almost Blank Subframe (ABS) muting for interference management in LTE-Advanced heterogeneous networks. This implies that more recent UEs (Release 10 and later) that use ABS muting can live alongside older UEs (Release 8).

While the more recent UEs can benefit from the improved interference coordination techniques, the older UEs do not require modification to support ABS. LTE-Advanced (Release 10 and later) uses ABS to reduce interference between various cell layers (such as macro and pico cells) in heterogeneous networks. In ABS, some subframes (time slots) are muted in one cell layer and used by the other layer. Because of this, certain frequent channels are still broadcast even while ABSs are fully muted, including Common Reference Signals (CRS) and, System Information Block (SIB) 1, paging, positioning reference signal, master information block, synchronization channels, and the PDCCH related to SIB1/Paging, in addition to Channel State Information Reference Signals (CSI-RS).

When ABS is used on a regular basis, time-variant interference variations are inevitable. For instance, the channel quality of a UE serviced by a pico will vary based on whether the macro is using ABS or regular transmission. The quality of a macro-UE near a CSG Femto will likewise differ based on whether the femto transmits standard subframes or ABS. Thus, UEs can be configured with time-domain measurement limitations: Channel State Information (CSI), which includes Rank Indicator (RI), Precoding Matrix Information (PMI), and Channel Quality Information (CQI). Depending on whether conflicting cells employ ABS or regular transmission,

2.6.3 Dynamic Interference Coordination in the Power Domain

Dynamic interference coordination in the frequency or time domain can be combined with updated power control rules to mitigate the negative impact of interference in various installations. For example, interference problem, where a small cell experiences elevated levels of uplink interference from a terminal allocated to a bigger cell, can be mitigated by adaptively reducing the terminal transmit power. The uplink interference that several femto UEs produce to nearby macro cells can also be lessened. In both cases, the UE transmission power rule can be written as follows:

$$P = 10\log_{10}.(PRBs) + P_0 + \alpha.PL + \Delta MCS + \text{Closed loop}$$

- PRBs denotes the number of scheduled Physical Resource Blocks (PRB),
- P_0 is the normalized transmit power density,
- α is the pathloss compensation factor,
- PL is pathloss, and
- ΔMCS is the quality requirement for the modulation and coding scheme chosen.

If α equal to 1, the power control fully compensates for the path loss. If α is less than one, we employ fractional power control. To optimize uplink performance for co-channel macro and pico situations, it's recommended to employ various P_0 settings for the macro and pico layers, as well as varied α values.

As a general guideline, the offset in P_0 between the two layers should be on the same order of magnitude as the difference in downlink transmit power levels between macro and pico, with the pico-layer having a smaller offset. However, in practice, the ideal choice of uplink power management settings is determined by a variety of circumstances and must be adjusted accordingly. Options include leveraging the closed loop component of the power control mechanism and auto-tuning of power control settings to maximize the desired key performance indicators for the uplink. In general, power control modifications always result in a trade-off between the performance of one cell and the other, thus the interference coordination parameters will also be determined by the mobile network operator's varied considerations.

2.6.3 DSS approaches

2.6.3.1 Type 1 DSS Approach

Type 1 – DSS approach has a single-transceiver single-band enabled SBSs operating in the licensed spectrums of homogeneous systems (Fig. 2.9): In this kind of SBS, the MBS and an SBS both use their own MNO spectrum while adhering to the MNO's interference control guidelines. Using the eICIC (Enhanced Intercell Interference Coordination) approach, this type of SBS can be used for the CSA (Co-channel Shared Access) between an MBS and an MNO's in-building SBSs. Although this sharing can lead to better coverage and better use of resources, it also calls for co-channel interference management techniques.

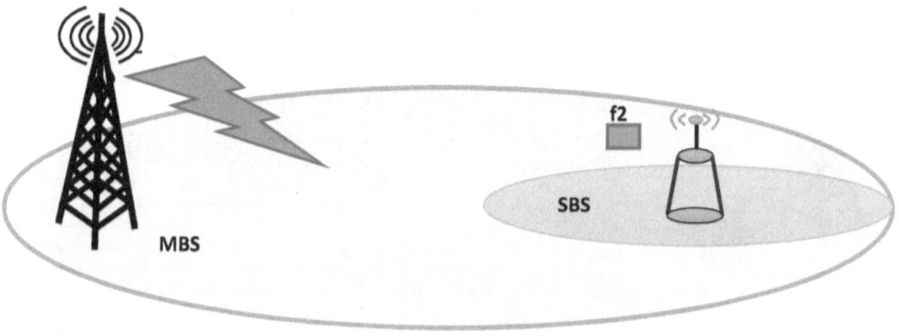

Figure 2.9 Type 1 SBS architecture.

Type 1 Small Base Station (SBS) heterogeneous networks involve deploying small cells alongside larger macro base stations (MBS) to enhance network performance and capacity. These networks are characterized by the integration of multiple SBSs within the coverage area of an MBS, creating a multi-tier architecture. Users can connect to either an MBS or an SBS, depending on the signal strength and network load.

NB: f_2 denotes the spectrum of the MBS.

2.6.3.2 Type 2 DSS Approach

Type 02 - DSS approach has a single-transceiver single-band enabled SBSs operating in the licensed spectrums of heterogeneous systems (Fig. 2.10): The MBS and an in-building SBS of an MNO operate at different frequencies. This Type of SBS can be used for the LSA (Licensed Shared Access) for sharing the licensed spectrum of a heterogeneous system with in-building SBSs of an MNO.

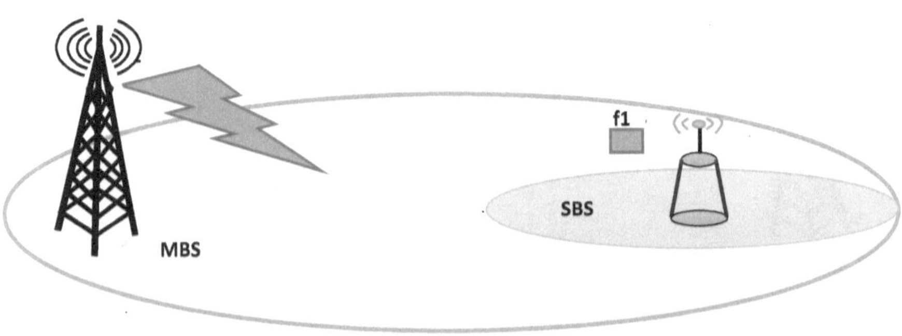

Figure 2.10 Type 2 SBS architecture.

MBSs and Pico Base Stations (PBSs) often operate in lower frequency bands like 2 GHz, while in-building SBSs operate in higher frequency bands like 28 GHz (mmWave).

NB: f_2 denotes the spectrum of the MBS.

2.6.3.3 Type 3 DSS Approach

Type 03 – is a single-transceiver single-band enabled SBSs operating in the unlicensed spectrums of heterogeneous systems (Fig. 2.11): Single transceiver, single-band SBSs operating in unlicensed spectrums within heterogeneous systems (SSU-HET) are a particular sort of small cell architecture. In this configuration, the macrocell base station (MBS) of the mobile network operator (MNO) and the small cell base station (SBS) use distinct frequencies, with the SBS using unlicensed spectrums. This technique meets the demand for higher capacity and coverage in indoor settings. This type of SBS allows the ULA (Unlicensed Shared Access) to run an MNO's in-building SBSs on an unlicensed heterogeneous spectrum band.

NB: f2 denotes the spectrum of the MBS.

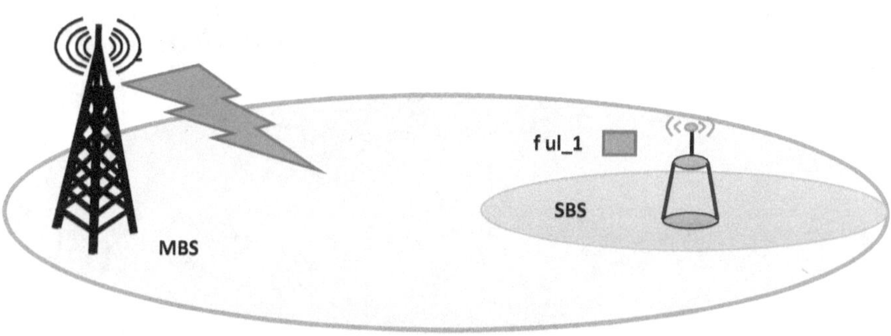

Figure 2.11 Type 3 SBS architecture.

2.6.3.4 Type 4 DSS Approach

Type 04 – is a single-transceiver multiband enabled SBSs operating in the licensed spectrums of homogeneous systems (Fig. 2.12): Single-transceiver multiband enabled Small Base Stations (SBSs) operating in the licensed spectrums of homogeneous systems are a type of SBS architecture that allows for indoor small cell operation on the spectrums of multiple Mobile Network Operators (MNOs) in the same area using a single transceiver. An in-building SBS operates in the spectrum of multiple MNOs in the same region using a single-transceiver by employing techniques such as carrier aggregation. The Authorized Shared Access (ASA), can be realized subject to interference management among MNOs.

NB: f_2 denotes the spectrum of the MBS.

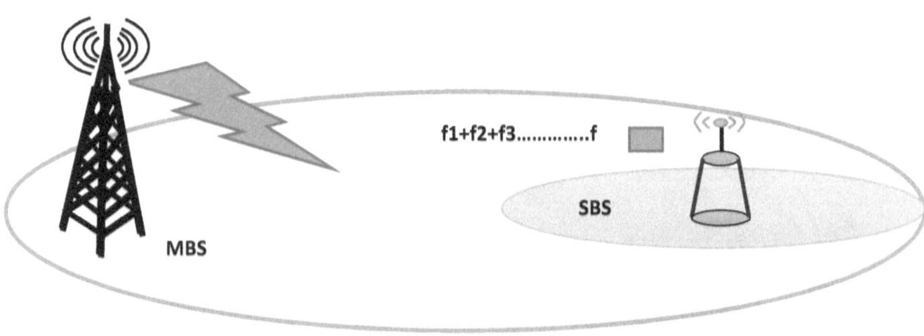

Figure 2.12 Type 4 SBS architecture.

2.6.3.5 Type 5 DSS Approach

Type 05 - multi-transceiver multiband enabled SBSs operating in the licensed spectrums of homogeneous systems (Fig. 2.13): Multi-transceiver, multi-band small base stations (SBSs) operating in homogeneous systems' licensed spectrums employ several transceivers to gain access to multiple spectrum bands within a single network operator's approved spectrum. This technology is commonly used to enhance capacity and meet the growing need for indoor data traffic in indoor environments, particularly buildings. Multiple transceivers are needed because the signal propagation characteristics at MNO operational spectrums differ. Spectrum pooling and spectrum rental are two types of Co-primary Shared Access (CoPSA) that can be used.

NB: f_2 denotes the spectrum of the MBS.

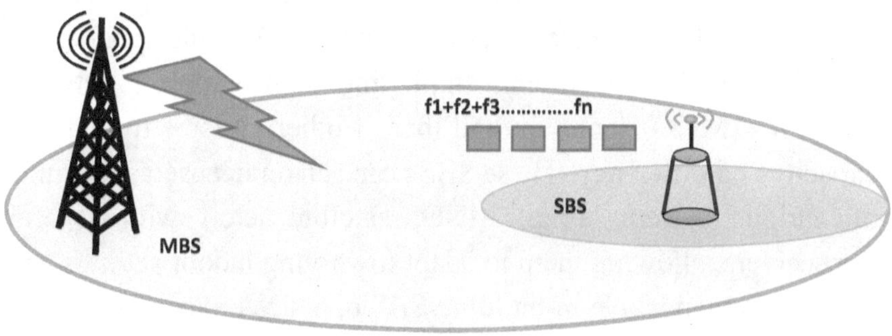

Figure 2.13 Type 5 SBS architecture.

LTE Co-primary Shared Access (CoPSA) is a spectrum sharing mechanism in which different Mobile Network Operators (MNOs) share their licensed spectrum while providing equal access rights to consumers. This is often conducted by agreements between operators that allow for either spectrum pooling, in which all operators share the full pool, or mutual renting, in which one operator can temporarily lease spectrum from another. CoPSA is intended to improve spectral efficiency and capacity, particularly in heterogeneous networks (HetNet's) having tiny cells. Instead of operating in isolation, MNOs pool or lease their resources to meet capacity demands and improve overall network performance. CoPSA relies on operator agreements to set spectrum sharing terms, such as access time and shared resources.

2.6.3.6 Type 6 DSS Approach

Type 06 - multi-transceiver multiband enabled SBSs operating in the licensed spectrums of heterogeneous systems (Fig. 2.14): Multi-transceivers, multi-band SBSs that operate in the licensed spectrums of heterogeneous systems provide a flexible approach to indoor network deployment, employing both a mobile network operator's (MNO) spectrum and that of other systems to increase capacity and efficiency. These SBSs can run on licensed spectrums of various systems (e.g., MNO, satellite, etc.) with several transceivers, allowing them to adapt to varying indoor settings and network demands. An in-building SBS operates on its own MNO's spectrum as well as a heterogeneous system (e.g., a satellite system) with various transceivers. The CSA can be implemented with one transceiver, while the LSA can be realized with another SBS transceiver.

LTE Co-channel Shared Access (CSA)describes a scenario in which both macro and low-power nodes in a heterogeneous network (HetNet) use the same frequency channels for uplink and downlink communications. While this strategy aims to improve spectral efficiency by allowing for greater frequency reuse, it may result in increased interference between network layers. LTE HetNet's consist of a variety of cell types, including macro cells (big coverage) and tiny cells (local coverage).

NB: f_2 denotes the spectrum of the MBS.

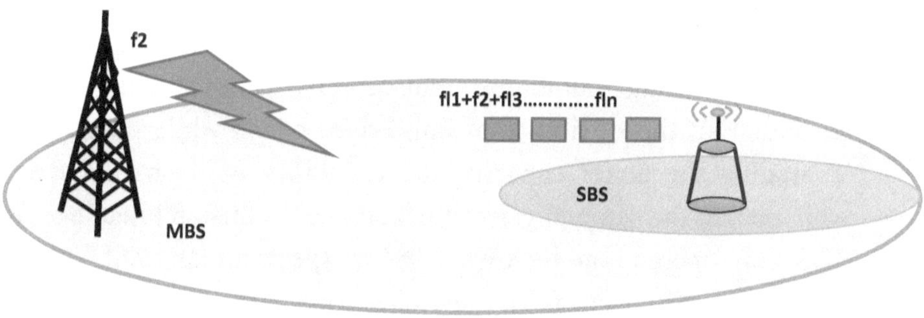

Figure 2.14 Type 6 SBS architecture.

Despite the limitations, co-channel sharing is sometimes preferred over exclusive spectrum allotment because it has the potential for increased spectral efficiency, particularly in dense deployments.

LTE's Licensed Shared Access (LSA) is a regulatory framework that allows mobile network operators (MNOs) to dynamically share licensed spectrum with other users, including incumbent users, in a controlled manner. This technique seeks to maximise spectrum utilisation by allowing MNOs to utilise temporarily available frequency bands, hence increasing network capacity and performance while safeguarding the rights of existing spectrum users. A database is used that keeps data on incumbent users' spectrum consumption and availability, allowing the LSA Controller to make more informed decisions about spectrum allocation.

2.6.3.7 Type 7 DSS Approach

Type 07 - multi-transceiver multiband enabled SBSs operating in the unlicensed spectrums of heterogeneous systems (Fig. 2.15): Multi- transceiver, multi-band Small Base Stations (SBSs) operating in unlicensed spectrums inside heterogeneous systems are a type of SBS design that allows a single SBS to communicate across several frequency bands, including licensed and unlicensed spectrums, using numerous transceivers. This strategy takes advantage of the flexibility of unlicensed bands while potentially leveraging the better capacity and reliability of licensed bands, whilst using the same physical SBS device. In this architecture, an SBS can operate on its own MNO's spectrum as well as the unlicensed spectrum of a heterogeneous system with various transceivers. The Co-Channel Shared Access (CSA), can be implemented using one transceiver, while the Licensed Assisted Access (LAA), can be implemented using another SBS transceiver.

NB: f_2 denotes the spectrum of the MBS.

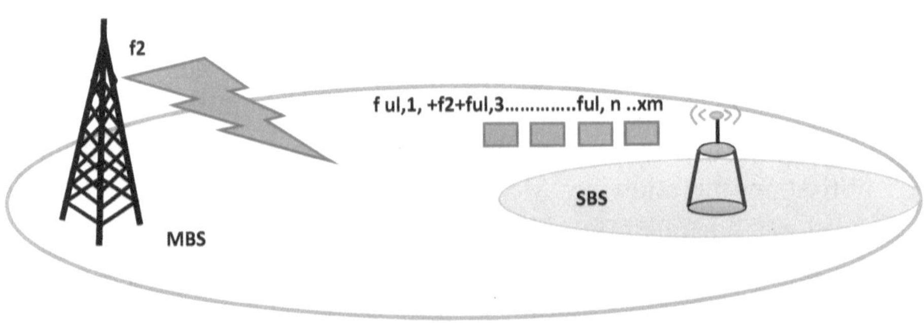

Figure 2.15 Type 7 SBS architecture.

LTE Co-Channel Shared Access (CSA) is the practice of allowing multiple LTE networks or devices to communicate on the same frequency channel, also known as co-channel deployment. This approach is often used to maximise spectrum utilisation, particularly in places with limited available bandwidth or where numerous operators use the same frequency. Successful co-channel sharing in LTE requires effective interference management.

Licensed Assisted Access (LAA) is a technology that enables LTE networks to use unlicensed spectrum (such as the 5 GHz band) alongside licensed spectrum to increase data capacity and throughput. It allows LTE to coexist with other technologies, such as Wi-Fi, in unlicensed bands using processes such as listen-before-talk (LBT). LAA is meant to cohabit with other unlicensed band technologies, including Wi-Fi, by utilising methods such as LBT and changing transmission characteristics. LAA, which uses unlicensed spectrum, can enhance data rates and network capacity, especially in congested locations. LTE-LBT (Listen-Before-Talk) is a method used in Long Term Evolution (LTE) systems to ensure equitable access to unlicensed spectrum alongside other technologies such as Wi-Fi. It is a channel access approach in which devices "listen" to the channel before broadcasting to avoid collisions and maintain coexistence with other users.

2.6.3.8 Type 8 DSS Approach

Type 08 - multi-transceiver multiband enabled SBSs operating in the licensed and unlicensed spectrums of heterogeneous systems (Fig. 2.16): These networks provide a means to increase capacity and spectral efficiency in indoor mobile networks. By allowing SBSs to use several bands and transceivers, as well as share both licensed and unlicensed spectrum, these systems can meet the demands of high data traffic and network density. One of an SBS's transceivers operates in a heterogeneous system's licensed spectrum (e.g., a satellite system), while the second transceiver runs in an unlicensed spectrum (e.g., 60 GHz unlicensed spectrum) via twin transceivers. Licensed Shared Access (LSA) can be achieved with a single transceiver, while Licensed Assisted Access (LAA) requires additional transceiver.

NB: f_2 denotes the spectrum of the MBS.

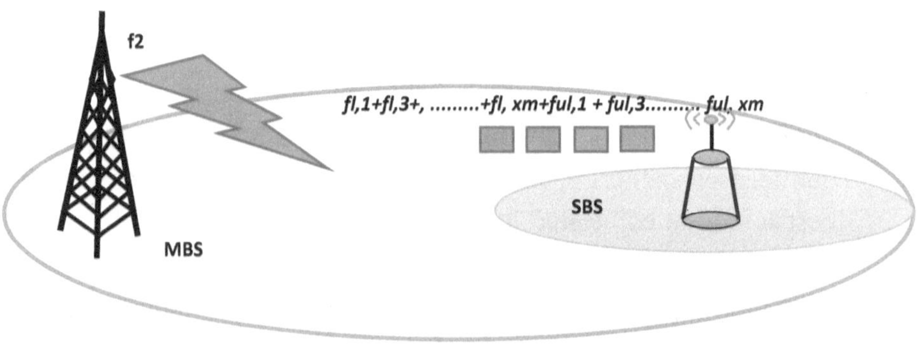

Figure 2.16 Type 8 SBS architecture.

2.6.3.8 Type 9 DSS Approach

Type 09 - multi-transceiver multiband enabled SBSs operating in the licensed and unlicensed spectrums (option 2) (Fig. 2.16):

NB: f_2 denotes the spectrum of the MBS.

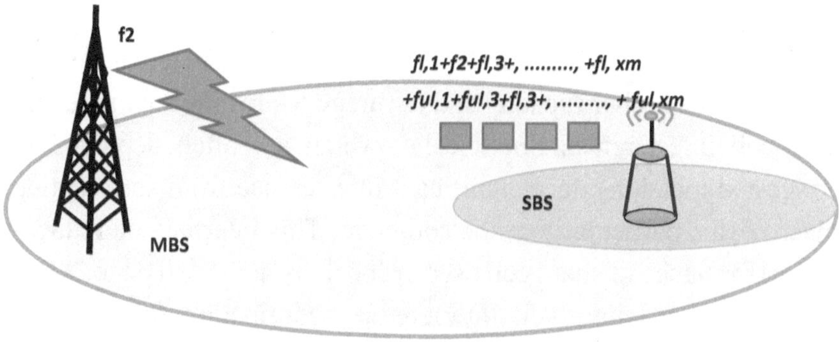

Figure 2.16 Type 9 SBS architecture.

Within this architecture, one of SBS's transceivers utilizes its own MNO's spectrum, another uses the licensed spectrum of a heterogeneous system (such as a satellite system), and a third employs numerous transceivers to operate on an unlicensed spectrum (such as the 60 GHz unlicensed spectrum). Transceivers 1 and 2 of the SBS can achieve LSA, transceivers 1 and 3 of the SBS can achieve LAA, and transceiver 1 of an SBS and the spectrum of its MNO can achieve CSA. Table 2 covers DSS approach implemented using an SBS architecture.

Table 3 Realized spectrum sharing accesses SBS architectures

SBS architecture	Type 1	Type 2	Type 3	Type 4	Type 5	Type 6	Type 7	Type 8	Type 9
	CSA	LSA	ULA	ASA	CoPSA	CSA	CSA	LSA	CSA,
DSS						LSA	LAA	LAA	LSA,
									LAA

2.6.1.1 ABS-based eICIC technique

Heterogeneous networks (HetNet's), which include macro and small cells, provide higher capacity and coverage while introducing severe inter-cell interference. Small cells, which are often deployed in congested locations, face significant interference from surrounding macro cells, particularly at the cell edge. This interference reduces the performance of small cell users, especially at the cell edge. ABS-based enhanced inter-cell interference coordination (eICIC) is a technique used in heterogeneous networks (HetNet's) to reduce interference between macro and small cells through the use of "almost blank subframes" (ABS). It entails marking some subframes in the macro cell as "almost blank," which means they have little or no transmission, allowing tiny cells to transmit without encountering considerable interference from the macro cell during those time periods.

Depending on its operating spectrum, the ABS-based eICIC technique is applied to any SBS transceiver to prevent CCI when sharing the licensed spectrum of a homogeneous or heterogeneous system.

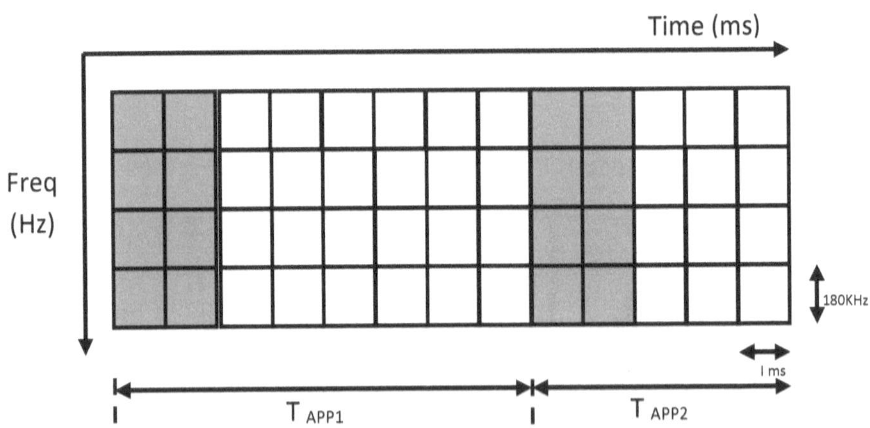

Figure 2.17 eICIC method based on ABS.

As illustrated in Fig. 2.17, the eICIC technique is founded on the idea that "an SBS architecture can be configured such that it can operate only during non-ABSs per ABS Pattern Period (APP)".

2.7 Spectral Efficiency for Capacity Enhancement

2.7.1 Interweave Cognitive Radio

Interweave cognitive radio (CR) is a spectrum access approach in which secondary users (SU), also known as cognitive users, take advantage of the radio frequency spectrum while it is not in use by main users (PU), also known as licensed users. This method use spectrum sensing to detect spectrum gaps, which are underutilised areas of the spectrum, and then uses those frequencies for internal communication. The method by which CR users determine the presence or absence of main users. This is critical for determining available spectrum holes. CR users broadcast only when the primary user is not using the spectrum, allowing for the best possible spectrum utilisation.

Due to the radio spectrum constraint, MNOs are having a tough time meeting the increasing consumer demand for high data speeds and capacity. In this way, spectrum access models have a significant impact on how well the finite amount of spectrum may be used. Government organisations generally allocate large sections of licensed spectrum to a country's MNOs over an extended period using the static access approach. Nonetheless, a significant disadvantage of the static access model is that, due to the exclusive right to use the allotted spectrum, a significant portion of the licensed spectrum may remain unused in terms of time, frequency, and space, while the other MNO may be unable to obtain enough spectrum, resulting in inefficient nationwide spectrum utilisation.

In recent years, cognitive radio has emerged as a crucial strategy for addressing spectrum underutilisation. Secondary users, or unlicensed devices, can only use the radio spectrum when the primary users, or licensed devices, are not in use, thanks to interwoven cognitive radio. To avoid interference with primary users, it uses spectrum sensing to identify available channels and dynamically adjust transmission parameters. The purpose of this method is to expand spectrum consumption without causing disturbance to main customers. One of the key aspects of Cognitive Radio is the technology that prevents Primary and Secondary UEs from clashing and allows one MNO to share its licensed spectrum with another to enhance effective spectrum bandwidth and support high data speeds and capacity. Cognitive Radio systems use three types of spectrum access techniques based on how Primary and Secondary UEs avoid collisions when accessing any frequency: interweave, underlay, and overlay. Consumers are disturbed.

Secondary UEs can only opportunistically access spectrum that Primary UEs do not use during normal access. The interweave model's main notion is to take advantage of existing underutilised spectrum and repurpose it in an opportunistic manner at minimum or no cost. More specifically, under the interweave paradigm, Secondary UEs can opportunistically use the unused spectrum in time, frequency, and geographic location of licensed Primary UEs on a dynamic shared-use basis without interfering with Primary UEs, such as when Primary UEs are inactive [11]. Standardisation organisations have endorsed the interwoven cognitive radio model because of its potential for addressing spectrum underutilisation and its ability to give acceptable dependability and guaranteed Quality of Service (QoS). In addition, Secondary UEs are allowed to transmit at full power even when interwoven access requires Secondary UEs to perform additional spectrum sensing to discover an idle spectrum of Primary UEs.

2.7.1 Interweave Strategy Based Shared-Use (ISSU) model.

To prevent co-channelling, the ISSU model considers sharing the entire mmWave spectrum of one MNO. To prevent co-channel interference between UEs of p-MNO and sMNO, the licensed mmWave spectrum of one MNO, i.e., Primary-MNO (pMNO), can be permitted to share with small cells in a building of another MNO, i.e., Secondary-MNO (s-MNO), only if no UE of p-MNO is present inside the corresponding building of small cells of s-MNO. This is the Interweave Strategy Based Shared-Use (ISSU) model for the dynamic spectrum access of licensed 28 GHz mmWave spectrum of one MNO to another. Otherwise, in-building tiny cells of s-MNO cannot share any spectrum with p-MNO.

Consequently, an s-MNO can use the interweave technique to opportunistically access the whole licensed spectrum of all p-MNOs in a country in terms of time, frequency, and space, provided that no UE of the corresponding p-MNO is present simultaneously within the same building of small cells of s-MNO. By doing this, an s-MNO continuously senses whether the shared spectrum is being used by any p-MNO in the building and instantly camps on a p-MNO's shared spectrum when a UE of the associated p-MNO is not there to run the s-MNO's small cells inside the building.

2.7.2 Underlay Cognitive Radio Spectrum Access (UCRSA)

In Underlay Cognitive Radio Spectrum Access (UCRSA), Secondary UEs can concurrently access Primary UEs ' spectrum at a lower transmission power to serve their users in the underlay spectrum access, if they meet the interference threshold that Primary UEs have set. Several factors are utilized by the underlay spectrum access technique. For instance, an SU can function concurrently with a Primary UEs at its spectrum, negating the necessity for the intricate spectrum detecting process in Cognitive Radio systems [12]. To identify an idle spectrum of Primary UEs, no additional switching is required. Furthermore, this contributes to the potential for transmissions without interruptions. The underlay access is also right for short-range communications since it restricts the transmission power of a Secondary UEs to the interference threshold that a Primary UEs has set up.

In this way, a small cell of an s-MNO can use its licensed 28 GHz mmWave spectrum at its maximum transmission power using the proposed approach, while a small cell of any p-MNO can use its shared 28 GHz mmWave spectrum at a lower transmission power at any time that corresponds to the predefined interference threshold set by the p-MNO to limit CCI to any UE of the p-MNO. The suggested method does not require tiny cells of an s-MNO to continuously sense to identify the existence of UEs of any p-MNO to share with the p-MNO's spectrum at any moment, like the traditional underlay spectrum access.

The combination of these two spectrum accesses can optimize the Spectral Efficiency (SE) and Energy Efficiency (EE), even though interweave and underlay each have advantages and disadvantages as previously mentioned. Secondary UE (Sus) can examine underlay access when the Primary UE (PU) spectrum is busy and interweave access when the PU spectrum is idle. In contrast to operating at lower power when using only underlay access all the time, this enables SUs to operate at their maximum power during an idle period.

Thus, the approach utilizes both underlay and interweave access strategies. In particular, as long as no UE of the corresponding p-MNO is present at the same time, small cells of an s-MNO in a building can use the interweave access to opportunistically access the entire licensed 28 GHz spectrum of every single p-MNO in a nation in time, frequency, and space by running them at their maximum transmission power. To restrict (CCI) Co-channel interference to the UE of the p-MNO, each small cell of the s-MNO instantly lowers its transmission power in accordance with the predetermined interference threshold set by the p-MNO if a UE of any p-MNO is present after the underlay access. To guarantee that the CCI constraint to UEs of the corresponding p-MNO is guaranteed, the small cells of the s-MNO continuously sense the existence of UEs for each p-MNO to update the corresponding spectrum access mode of operation to either the interweave access or the underlay access.

2.8 Traffic Steering for Capacity Enhancement

Based on several variables, LTE Traffic Steering is a strategy that dynamically routes user traffic to various cells, frequency bands, or radio access technologies (RATs) to maximise network capacity and performance. The objective is to enhance user experience, boost network capacity, and lessen congestion by utilising the capabilities of various network resources. By directing traffic to the RAT or tier that provides the best performance in terms of any system of measurement of interest, LTE and advanced LTE traffic steering allow mobile network operators to maximise resource utilisation, maximise customer perception, and minimise power consumption at the base station or terminal. This need, which ensures reliable and optimal mobility performance including a manageable number of handovers and the avoidance of ping-pongs and radio link failures which is directly tied to mobility management.

Traffic steering is essential for reducing operating expenses (OPEX) and allowing for the management or postponement of capital expenditures (CAPEX) in heterogeneous networks. Offloading as much traffic as possible to WiFi or femto cells is a frequent strategy in this scenario, as they often require a low-cost backhaul infrastructure. Furthermore, more efficient use of base station infrastructure is made possible by load balancing between RATs and the other layers, which raises return on investment and offers a more reliable quality of service.

When assessing traffic steering, the following factors need to be considered: network and terminal capabilities, required services and quality of service (QoS), load in various RATs and layers, base station and terminal power consumption, traffic steering, signal strength and interference, and terminal speed. The many traffic steering strategies that LTE and LTE-Advanced provide are frequently separated into those that apply to connected and idle terminals. While load balancing and ensuring uninterrupted service provision are the goals in connected state, the goal in idle state is typically to allow a terminal camp on the RAT and the layer it may be supplied with once it forms a connection in the future.

2.8.1 Traffic Steering in Idle State

To maximise network performance and resource utilisation, traffic steering in the LTE idle state involves steering User Equipment (UE) to various cells or layers according to variables including load, radio circumstances, and user preferences. This is conducted by modifying the Cell Reselection process's parameters, which affect which cell the UE will camp on when it is not actively sending data. When UEs are in the right place and the pico cell has capacity, traffic steering can be conducted by directing them to the pico cells in a multi-layer network that has both macro and pico cells. By doing this, the macro cell's traffic is offloaded, enhancing both the macro and pico layers' performance. Advantages of Idle State Traffic Steering:

- ***Better Load Balancing*:** More evenly distributes traffic throughout the network.
- ***Improved network capacity***: Steering in idle state makes it possible to make better use of the resources that are available.

- ***Decreased Handovers:*** The frequency of handovers during active sessions can be reduced by first choosing the most suitable cell.
- ***Battery Savings:*** Less frequent measurements may result in less power being used.

The terminal itself oversees determining the RAT or frequency layer and sensing its surroundings while it is in idle state, which is when it has no radio resource control link to the radio network. However, by giving the terminal distinctive characteristics within so-called system information blocks (SIBs), like thresholds and priorities involved in the cell reselection process, the network can affect the terminal's choice.

System information blocks (SIBs) are essential to cellular networks, especially Long-Term Evolution (LTE) and 5G networks. In SIBs, the base station (eNodeB in LTE or gNodeB in 5G) broadcasts important system-related data to mobile devices (UEs) within its coverage area. In many diverse kinds of network-related operations, including network access, handover, initial cell selection, and cell reselection, these informational blocks assist UEs. SIBs are transmitted over broadcast channels (BCH), which are specialized control channels that are always in operation. Valuable information is contained in the BCH that UEs need to join to the network. SIBs are regularly distributed to UEs so they can update their system settings and make educated judgments regarding network access and mobility. SIBs are sent as part of the larger System Information (SI), which includes vital data such as neighbour cell details, scheduling information, and cell parameters. Although the fundamental idea behind SIBs is the same for all technologies, 5G NR adds optimisations to cut down on overhead and boost productivity. The base station regularly transmits these SIBs, which UEs can decode and extract the pertinent data from. To facilitate effective network access, mobility, and handover processes, the SIBs give UEs the necessary information to connect to and stay connected to the cellular network. The following list explains the standardized cell reselection processes in LTE that can be applied to traffic steering:

2.8.1.1 Traffic Steering Techniques in Idle Mode

In true terms, a Traffic Steering (TS) policy entails implementing user distribution throughout the layers to achieve maximum utilisation of resources in accordance with the policy. Due to the signalling and data transfer costs associated with HO; it is crucial to attain the ideal user distribution from the idle state. By fully aligning user distributions in both connected and idle modes, the ping-pong effect between layers during mode changes is likewise prevented.

Cell selection is the process of choosing a cell after turning on the UE, cell reselection is the process of choosing a cell because of mobility, and the cell to which the user is attached in idle mode is known as the camped cell. In idle mode, the user distribution can be changed by a variety of methods [13]. For example, a feature that gives large and small cells varied priorities for cell reselection is a hierarchical cell structure. To prevent constant reselections, this feature aims to give preference to mobiles with high mobility when they reselect towards large cells. Because of this, this functionality would be restricted in a more diverse setting where a layer is not solely based on cell size and where various goals, outside of those pertaining to high-speed users, may exist. Another feature that operators can use to bar a cell that the UE cannot camp on is cell barring; however, this strict approach may result in less-than-ideal performance. Basic biassing is an additional strategy that involves modifying certain offsets in the cell reselection process to alter the cells' attractiveness for selection.

This implies that the range of the region where cell edge users are likely to choose the cell can be increased or decreased in the intra-frequency situation. However, from the Traffic Steering (TS) point of view, basic biassing is limited to cells that belong to the same RAT in modern technologies like LTE. Different RATs and complete overlap of the coverage zones are possible in HetNet scenarios. In this situation, the absolute priority technique becomes an effective way to direct users, whether they are cell edge users, towards a different RAT or frequency. This approach relies on giving the network layers varying priorities such that the user distribution in idle mode can be effectively changed by only altering the layers' priorities.

2.8.1.2 Absolute priorities

"Absolute priorities" in LTE cell reselection means that a UE (User Equipment) gives certain cells more importance than others based on preset priority levels. The network usually sends these levels through system information or RRC (Radio Resource Control) messages. These priorities are particularly important for figuring out which cell a UE will camp on when it is idle or inactive. They also affect how quickly a device can move between cells. Absolute priorities are given numbers, usually between 0 and 7, with higher numbers meaning higher priority.

A UE continuously assesses nearby cells according to signal strength and priority when it is in idle mode. The UE may reselect to a higher priority cell (or a lower priority cell that satisfies specific requirements) if it is detected, which could prolong battery life or improve connection quality. System Information Blocks (SIB), such as SIB3, SIB4, SIB5, SIB6, and SIB7, are frequently used to broadcast priorities for various frequencies and RATs (Radio Access Technologies). In certain situations, the network may use the RRC Connection Release message to indicate priorities for UEs. When it comes to controlling mobility between various frequencies (like moving from 2G to 4G) and RATs (like moving from LTE to

WCDMA), absolute priorities are especially crucial.

LTE employs standards, like priority comparisons and signal strength thresholds, for cell reselection. UE camped on a WCDMA cell, for example. The UE will reselect the LTE cell if it detects a nearby LTE cell with a higher priority (and satisfies the signal strength threshold). Higher priority cells will be given priority by the UE, which means it will attempt to remain camped on them for as long as the signal quality is satisfactory. Based on pre-established rules and network configurations, the network directs UEs towards the most appropriate cells with the aid of absolute priorities in LTE cell reselection. This guarantees that electronic devices can effectively control their connections, maximize battery life, and keep a dependable connection when switching between cells. In addition, the network can specify absolute priorities for various intra- or inter-RAT carriers, which are then sent to terminals either through broadcast or connection release. The latter choice is especially intriguing since it enables the individual definition of absolute priorities for every terminal. To fully utilize the new LTE radio network capabilities, LTE radio could be given a higher priority.

2.8.1.3 Absolute Priority-Based Cell Reselection Algorithm

The Absolute Priority-Based Cell Reselection Algorithm is a system used in mobile communication networks that allows a User Equipment (UE), such as a mobile phone, to transition between different cells (and possibly different Radio Access Technologies or RATs) when idle. This switching is based on predefined priorities and signal quality evaluations, which ensure that the UE connects to the most appropriate cell. In most cases, if one or more requirements are met, the UE automatically chooses a new cell to camp on. In this procedure, the cell ranking criterion R and the cell selection criterion S are typically assessed. The former ranks cells according to the signal quality or signal level, whereas the latter identifies candidate cells with signal quality (or level) over a minimal threshold.

The UE may be given a new set of parameters called absolute priorities (APs) for cell reselection in a setting with varying frequencies and RATs [14]. More precisely, during the cell reselection process, the operator can prioritise the inter-frequency and inter-RAT network layers thanks to the AP feature. When the reference signal received power (RSRP) from the source cell falls below a threshold and the RSRP from the target cell rises above another threshold, a UE device formally reselects towards a lower-priority cell.

On the other hand, when the Reference Signal Received Power (RSRP) from the higher-priority layer surpasses a threshold, a UE device reselects towards a higher-priority cell. Keep in mind that the latter scenario, where the UE switches to a higher priority, is less constrained because only one requirement that is, the target cell needs to be met. This is because shifting towards tiers of higher priorities is desirable. Furthermore, cells of higher priority will take precedence over a lower-priority frequency, or RAT, when multiple cells of different priorities meet the previously described requirements for reselection.

2.8.1.4 Adjustment of Absolute Priority for traffic steering

Both mobility parameters that is, the thresholds and the Absolute Priority (APs)can be adjusted in the AP-based cell reselection algorithm to conduct traffic steering (TS) strategies. To make the target cell more appealing, for example, the threshold that reselects towards higher priority could be lowered. But the Absolute Priority APs can also change how users are distributed across the layers without changing the steered users' signal level, which is usually a better option.

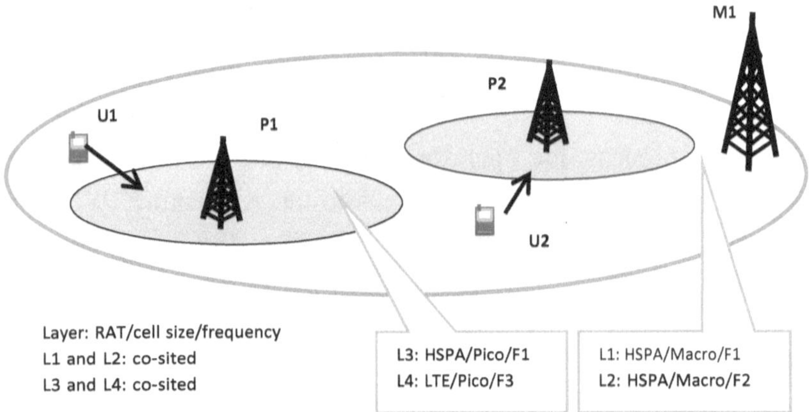

Figure 2.17 *Deployment scenario, composed of four network layers: two collocated macro layers and two collocated pico layers.*

A network architecture in which both macro and pico cell base stations are placed in proximity (collocated) within the same geographic area, forming a multi-layered network, is known as a four-layer network deployment scenario with two collocated macro layers and two collocated pico layers. By utilising the advantages of both macro and pico cells, this configuration is often utilised to increase network capacity and coverage. However, according to the 3GPP standard, there are certain restrictions when allocating priority levels to network layers.

Specifically, even though the layers have different cell sizes, they must have the same Absolute Priority (APs) if they have the same RAT and frequency. HSPA and LTE layers, for instance, must have separate Absolute Priority (APs) APs because it is prohibited for different RATs to share the same priority level. The operator's ability to assign APs to network layers is thereby reduced, which could potentially impact network performance. An example of how those limitations are considered, the performance impact is examined, and some suggestions for allocating priorities in intricate HetNet's scenarios are made in the paragraphs that follow.

Suppose that the HetNet illustrated in Figure 1 has been put into use. The four network layers that make up this deployment scenario each have many cells with comparable properties. In addition to L1, a high-speed packet access (HSPA) escape carrier (F2) has been implemented as a second macro layer (L2), which is usually utilised by operators to give macro users more capacity because it is not affected by other layers [8]. Assuming that base stations are situated in the same location, let us say that L1 and L2 are collocated layers. This configuration would theoretically be utilised in LTE deployment where LTE macro coverage is not considered.

However, LTE and HSPA Pico layers which are also collocated have been installed in crowded areas to increase capacity. APs must be assigned to the network layers based on these presumptions. Utilising the current picocells to their full potential by offloading as much traffic as possible from the macrocells would be a logical traffic steering (TS) strategy. But as has been pointed out, there are some restrictions to consider. First, the 3GPP states that since L1 and L3 have the same frequency (F1) and RAT (HSPA), they share the same AP.

Additionally, because the pico layers have distinct RATs, they cannot have the same priority. It would make sense to give L4 the greatest priority to unload the macrocells in accordance with the steering (TS) policy, as L1 interferes with L3's co-channel. The difference between them is in the relative priority of L1 and L2, i.e., whether the macro escape carrier (configuration indicated by AP1) takes precedence over the main carrier (AP2) or the opposite (AP2). In addition, take note of the fact that L1 typically has fewer UEs than L2 because of interference from L3, which has the same frequency as L1. By lowering the proportion of UE devices connected to the interfering layer (in this example, L1 and L3), APs can be utilised, for instance, to lessen the interference produced over other (intra-frequency) layers. Second, there is not much macro-offloading that happens when small cells are given more priority.

2.8.1.5 Cell Reselection Hysteresis

Certain cells can be prioritized over others due to the quality offset, while hysteresis prohibits repeated cell reselections caused by slight signal variations. They guarantee a reliable and effective cell selection procedure when combined. A "delay" in cell reselection is introduced by hysteresis. It makes sure a UE does not instantly move to a better cell when it notices a slight increase in signal strength. When the serving cell's signal strength is compared to that of nearby cells, a hysteresis value (such as Qhyst in WCDMA) is added. The UE will not reselect if the signal strength of the serving cell is still higher than that of the nearby cell after the hysteresis value has been added.

When a UE is camping on cell A and a nearby cell B has a marginally stronger signal, the UE will only switch to cell B if the signal strength of cell B is noticeably greater than that of cell A (taking the hysteresis value into account). Cell Reselection Hysteresis stops frequent cell reselections (the ping-pong effect) brought on by minute changes in the signal, which can be brought on by interference or fading, it specifies the priority of EUTRA-FA (Evolved Universal Terrestrial Radio Access-Foreign Area) during idle re-selection or mobility control information. LTE cell re-selection uses priority-based levels starting from 0 to 7 as per 3GPP 36.331 where Priority 0 is Lowest and Priority 7 is highest. Level hysteresis for cell reselection is applied when the new cells are in different location area. For instance, where a default value is 6dB; and Value ranges from 0dB to 14dB.Figure below gives an example of border cells. C1 controls which cell is originally selected, whereas CRH prevents rapid switching between cells during reselection, especially in border areas. The network employs

CRH to provide a steadier connection for the user. During the first cell selection process, a mobile station (MS) selects the cell with the greatest C1 value. A hysteresis value is added to the serving cell's C1 value during reselection, making it harder for the MS to switch to a neighboring cell unless the neighbor's C1 is significantly better.

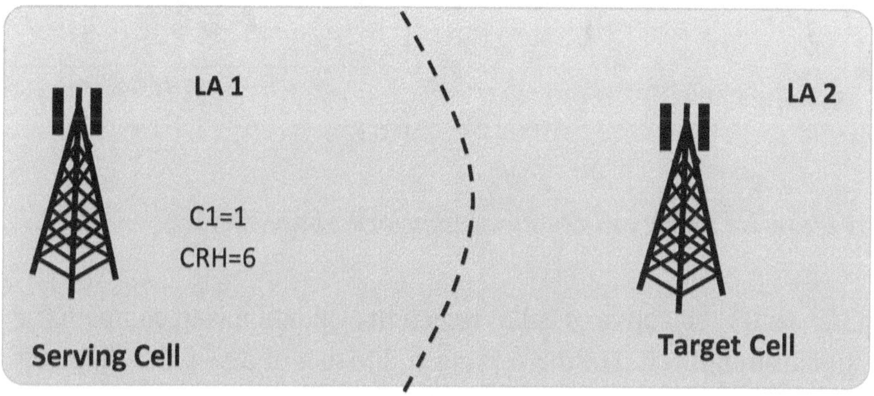

Figure 2.18 LTE cell re-selection priority-based levels starting from 0 to 7.

The mobile device may frequently switch between location zones if it is travelling in a location area. Every change need updating the location, which increases signalling load and increases the chance of paging message loss. The cell reselects hysteresis parameter CRH is used to stop this. Only if the cell's C1 is greater than the C1 of the serving cell by the Reselect Hysteresis value will the cell in the other location area be chosen. The CRH transmitted by the serving cell will be utilised for comparison because each cell may have a different CRH value. The device will ping-pong between location zones if the setting is set extremely low, increasing the signalling load. The mobile device may spend too much time in the incorrect cell if the value is set extremely high.

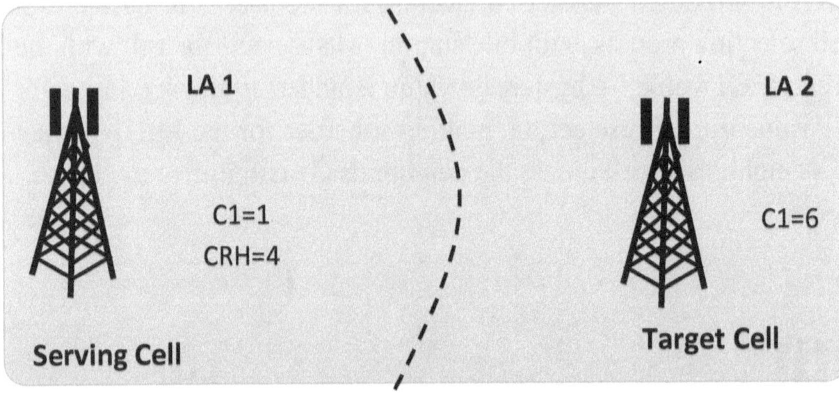

Figure 2.19 *LTE cell re-selection priority-based levels*

The central objective of cell reselection evaluation is to compare the signal strength (Rs) of the serving cell to that of a neighbouring cell (Rt). The following is a simplified representation:

Rs (Serving Cell) = Qmeas, s + Qhyst and Rt (Target Cell) = Qmeas, t – Qoffset.

Where:

- ***Qmeas, s:*** *Measured RSRP (Reference Signal Received Power) of the serving cell.*
- ***Qmeas, t:*** *Measured RSRP of the target cell.*
- ***Qhyst:*** *Hysteresis value for the serving cell (from SIB3).*
- ***Qoffset:*** *Cell- or frequency-specific offset (for inter-frequency cells).*

A UE will reselect to a target cell if its signal strength, including hysteresis, exceeds that of the serving cell for a set period (T_RESELECTION). The comparison is often as follows:

Rs < Rt, for a duration of T_RESELECTION, then reselect.

2.8.1.6 Quality Offset

The term "Quality Offset" in LTE refers to Reference Signal Received Quality (RSRQ) offset, a metric that is employed in choices about cell selection and handover. It affects the possibility that the user equipment (UE) will move to a nearby cell by adding a margin to its RSRQ. A cell's appeal for handover decreases with a greater offset. Quality offset enables network administrators to prefer specific cells or frequencies during cell reselection.

When comparing the serving cell's signal quality metric (e.g., RSRP or RxLev) to that of surrounding cells, a quality offset value is applied. If the offset is positive, it makes the serving cell appear better than it is, thereby delaying or preventing reselection to a neighbouring cell. If it is negative, it makes the serving cell appear worse, which increases the likelihood of reselection to that cell. In a network with many frequency bands, a higher priority frequency band may have a positive quality offset, increasing the likelihood that UEs will remain camped on that band, even if another band has slightly stronger signal strength. This metric enables network operators to regulate load distribution across multiple cells and frequencies, hence improving network performance and user experience. The fundamental reasoning underlying cell reselection using offsets can be summarised as follows:

Triggering condition:

$$MEAS\_neighbor + Qoffset\_neighbor, frequency - Hysteresis >. Threshold$$

Cancellation condition:

$$MEAS\_neighbor + Qoffset\_neighbor, frequency + hysteresis < threshold.$$

Where:

- ***MEAS_neighbor:*** *The neighbour cell's measured signal strength (for example, RSRP or RSRQ).*
- ***Qoffset_neighbor,frequency:*** *The quality offset value for the neighbour cell at the specified frequency.*
- ***Hysteresis*** *is a little margin added to prevent quick switching between cells.*
- ***Threshold:*** *The reselection threshold set by the system.*

2.8.1.5 Cell barring and block-listing

Cell barring and block-listing are methods used to control the selection and reselection of cells by User Equipment (UE) in idle mode to maximise resource usage and network performance. A prioritised list of cells to take into consideration is created via cell barring, which stops UEs from camping on particular cells, and blacklisting, which tells UEs to stay away from particular cells during reselection.

A cell may be blocked by the network, or it may offer a blacklist of cells that a terminal is not allowed to camp on. Blacklisting and cell barring are two different but related ideas that both involve limiting access to cellular services. Blacklisting is the process of preventing a device (identified by its IMEI) from connecting to any network because it has been reported lost, stolen, or linked to fraudulent activity. Cell barring, on the other hand, can be used to restrict which calls are permitted by networks or individuals. It can also refer to the barring of a particular cell (base station).

2.8.1.6 Cell Range Extension

Cell range extension (CRE) in idle mode in cellular networks, especially LTE, refers to methods which enables User Equipment (UE) to connect to a cell (usually a small cell such as a picocell or femtocell) even when it is on the outskirts of that cell's coverage region. To make smaller cells appear more enticing to UEs in idle mode, cell selection and reselection parameters are routinely modified, potentially employing strategies such as blank subframe utilisation. Low power nodes (LPNs), including relay nodes, femto cells, Pico cells, and RRUs/RRHs, are placed in a macro cell as part of a Heterogeneous Network (HetNet), which was first introduced in 3GPP Release 10. The main purposes of adding small cells are to fill in areas that are not served by the macro network, both indoors and out, and to boost capacity in hotspots with high user demand. By offloading from the massive macro-cells, they also enhance service quality and network performance. The result is a HetNet that offers higher bitrates per unit area by combining small cells with as big macro-cells.

Using so-called range extension basic biassing to make small cells larger is one way to direct more traffic towards smaller cells. This implies that, as shown in Figure 2.20, a terminal that really receives a better downlink signal from a nearby macro cell than from a small cell connects to the latter. Although range extension is a quite easy and efficient way to improve small cell offload, it must be utilised carefully since, as we will see later, too-strong biases can result in troublesome interference constellations.

Ensuring that the small cells truly serve enough customers is a significant challenge in HetNet planning. A positive cell selection offset to the small cell's received power can be used to expand the area it serves, which is one method of doing so. Cell range extension outspreads a cell's coverage by connecting a UE to a cell that is weaker than the strongest detected cell. The UE typically camps on the cell with the strongest received DL signal (SSDL) in a network where all cells use the same frequency (frequency reuse = 1). As a result, the border between two cells is situated at the place where SSDL is the same for both cells. This also usually corresponds to the point of equal path loss for the UL (PLUL) in both cells in homogeneous networks.

Compared to LPN, macro cells in a HetNet transmit at a significantly high power. There are two boundaries, as seen in the figure 2.20 below: an uplink cell border defined by PLUL and a downlink cell border defined by SSDL. At a point nearer the pico cell, which creates the Downlink cell border, the SSDL measured from the macro cell and the pico cell in the Downlink is equal. On the other hand, the pico cell is located far from the point where route loss to macro and pico cells is equal. This is because whether the UE is being supplied by a macro cell or a pico cell, the UE transmission power is the same.

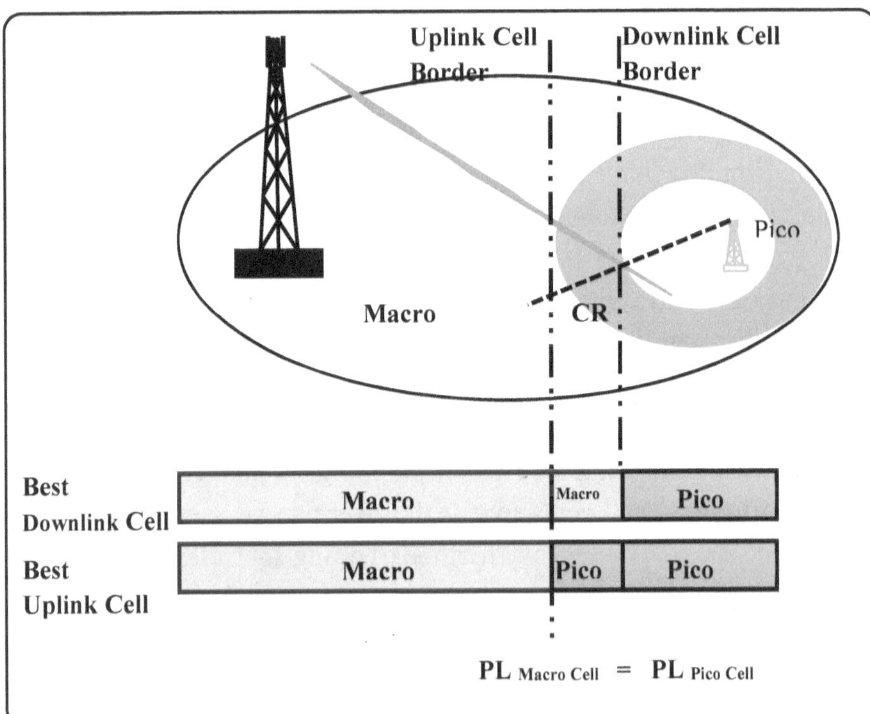

Figure 2.20 An uplink cell border defined by PLUL and a downlink cell border defined by SSDL.

Without the Cell range extension (CRE) function, the UE's choice of serving cell is determined by the power received by the DL. By adding a positive cell selection offset to the small cell's serving, secondary, and downlink (SSDL), Cell range extension (CRE) expands the small cell's service area. Small cells therefore draw more UEs than they would in the absence of Cell range extension (CRE), which significantly boosts HetNet efficiency. The UE is not always serviced by the eNB with the strongest DL received power in the Cell range extension (CRE) (shaded zone) of the preceding figure. Apart from boosting HetNet efficiency and reducing traffic from macro cells, Cell range extension (CRE) also offers the benefit of less uplink interference in the system.

The macro cell nearer the pico cell boundary would service a UE without CRE, which would interfere with the small cell's uplink by transmitting at a higher power. With CRE, a smaller, closer cell with less path loss is likely to service the same UE, lowering the need for the UE to transmit at such a high power. Implementing a selection offset (UEs in idle mode) or handover threshold in measurement configuration (UEs in connected mode) in favour of the small cell increases its range, as was previously described.

After covering CRE features, one of its main disadvantages is that, although the macro cell generates more downlink power than the small cell, the UEs are being forced fed by the small cell in the CRE zone. One of the most important metrics in wireless communication is SINR, or Signal-to-Interference-plus-Noise Ratio, which measures how well a signal is received. It shows the strength of the intended signal in relation to the sum of the strengths of background noise and interfering signals. Higher SINR readings often signify a higher-quality connection with reduced noise and interference. Therefore, if the SINR at the UEs in CRE that are supplied by small cells is less than 0 dB, there will be significant downlink inter-cell interference from the macro cell to UEs receiving transmissions from a pico cell. This could specifically affect how the DL control channels are received. The interference issue in HetNet's with small cells can be lessened by employing an assortment of capabilities that have been added to the 3GPP LTE specifications.

2.8.2 Traffic Steering in the Connected State

In the connected state, traffic steering and mobility management entail guiding user traffic to the best available network resources and guaranteeing uninterrupted connectivity when users are on the move. When users switch between cells or access points, mobility management manages the handover process, while traffic steering concentrates on maximising data flow across various network channels. These features are essential for preserving the best possible user experience and network performance.

The network side determines if a handover should be conducted between cells, RATs, or layers in the connected state; however, this decision is also based on the terminal's measurements. The frequency and extent of these measurements, nonetheless, can be influenced by the network. In LTE and LTE-Advanced, the network can send each terminal a list of carrier frequencies, from which the terminal can independently search cells based on predetermined measurement configurations. A black cell list, which lists the physical cell identities (IDs) that a terminal is not allowed to measure on specific carriers, is another way the network might limit the scope of these measurements.

Information interchange between base stations is obviously necessary for a well-tuned and system-wide mobility management. This exchange must be dependent on the X2 interface being available in LTE and LTE-Advanced. RAN Information Management (RIM) protocols may be used for the exchange in inter-RAT settings. To prevent ping-pong effects or contradictory policies, it is necessary to ensure that all traffic steering and mobility management tools utilised for idle and connected state terminals are properly aligned.

2.8.2.1 LTE Events Measurements Reporting

To provide smooth and seamless transitions between cells or different radio access technologies, ***Events Measurements Reporting*** are crucial for deciding when to start handover procedures in LTE networks. Depending on the needs and design of the network, these events may be used differently. The network evaluates the serving cell's reference signal received power (RSRP). If the RSRP exceeds the A1 threshold (for example, -85 dBm), the connection remains stable, and no mobility action is necessary. When RSRP goes below the A2 threshold (for example, -100 dBm), it signals low signal strength.

A network should ideally allow UE to report the signal quality (often RSRP) of both the target and serving cells and establish an arbitrary handover mechanism. However, since the network may require several consecutive measurement results rather than relying on just one or a few recorded signal quality values, this can be overly intricate and burden the infrastructure. The following is a brief explanation of 3GPP 36.331 5.5.4 Measurement report triggering. It would be considerably simpler and more natural to comprehend the nature of these occurrences if the detailed descriptions of each event's precise process were converted into a graphical representation.If the measured value crosses (rises or falls) a specific target value, a measurement report is generated.

There are two ways to set the goal value. Two methods are available: the first is the threshold, which is a type of absolute value; the second is the Offsef value, which is a type of relative value that refers to'something', such as serving cell value. In real situations though, the measured value could differ considerably , sometimes due to UE modem measurement errors and other times due to radio channel shifts in volatility. Generally speaking, the network would not be interested in such small variations and would prefer not to receive an excessive number of measurement reports that are produced by it.

The 'Hysterisis' parameter is provided in order to prove that this type of excessively frequent measurement report is produced by a tiny range of fluctuation. Perhaps the following depiction of the role of hysterysis makes obvious to you.The information that follows is a description of 3GPP 36.331. Measurement report triggering. It would be considerably simpler and more natural for someone to understand the nature of these occurrences if the detailed descriptions of each event's precise process were converted into a graphical representation.

Table 2.2 LTE HO triggering Events 3GPP 36.331

Event	**Description**
Event A1:	When the signal strength or quality of the serving cell exceeds a predetermined threshold, an event takes place. This signal, which is utilised in intra-frequency handovers, shows that the conditions of the serving cell have improved and may no longer require a handover. The purpose of this event is to discontinue measuring adjacent cells during inter-frequency handovers.
Event A2:	Event A2 is initiated when the serving cell's signal quality or strength drops below a predetermined level. It is employed in inter-frequency handovers and indicates that there may be a need for a handover since the serving cell's condition has become worse. This event is used to begin surveying nearby cells during inter-frequency handovers.
Event A3:	When a nearby cell's signal is better offset than the serving cell's, this event is triggered. Both intra-frequency and inter-frequency handovers employ it. Event A3's initiation indicates that the nearby cell's circumstances have considerably improved, making it a possible target for handover.

Event A4:	When the signal strength or quality of an adjacent cell surpasses a certain threshold during inter-frequency handovers, event A4 is triggered. This occurrence raises the possibility of a handover since it implies that the conditions in the neighbouring cell are better.
Event A5:	In inter-frequency handovers, event A5 is triggered when the signal from the serving cell falls below threshold 1 and the signal from a nearby cell rises beyond threshold 2. This occurrence suggests a possible handover since it shows that the circumstances in the serving cell have deteriorated while a nearby cell provides better signal conditions.
Event A6:	Event A6, which was first introduced in Release 10 for Carrier Aggregation (CA), is triggered when a neighbouring cell offsets better than the serving cell. The usage of several component carriers during intra-frequency handovers with carrier aggregation is what causes this event.
Event B1:	This event is used during inter-technology handovers (Inter-RAT). It occurs when an inter-RAT neighbour cell's signal exceeds a predetermined threshold. It implies that the inter-RAT neighbour cell's conditions are favourable for a possible handover.
Event B2:	In inter-technology handovers, event B2 is triggered when the serving cell's signal falls below threshold 1 and the signal of an inter-RAT neighbour cell rises above threshold 2. This event indicates that the serving cell's circumstances have deteriorated, whilst the inter-RAT neighbour cell delivers better signal conditions, implying a possible handover.

2.2 Recurrent Issues Regarding Measurement Reports

In 3GPP networks (LTE and NR), measurement reports are critical for Traffic Steering, mobility choices, and providing a great user experience. However, a network may have a number of issues due to improper measurement settings, environmental factors, or

a. **Inaccurate Measurements:** Numerous factors might result in inaccurate measurement reports, which frequently influence bad network choices. For instance, the UE may report erroneous signal strength (e.g., RSRP) or quality (e.g., RSRQ) values due to faulty network configurations, such as misaligned measurement offsets or unsuitable frequency configurations. This issue can also be caused by device-related problems, including a malfunctioning UE antenna or software errors in the measuring methods. Raw readings can potentially be distorted by environmental factors, such as multipath fading or interference from nearby cells. The network may make poor choices, such as switching to a lower-quality cell or failing to recognise a better neighbouring cell, if these errors are not reduced for example, by appropriate filtering (such as Layer 3 filtering) or calibration. Dropped calls, lower internet speeds, or a generally worse user experience can result from this.

b. **Delayed Reporting:** In situations involving high-speed mobility, like when a user is driving or riding in a train, delayed measurement reporting might lead to serious issues. The UE must promptly identify and notify any changes in signal conditions in these situations in order to guarantee timely handovers. However, a number of things can cause delays. For example, the UE may have a lengthy time-to-trigger (TTT) set in the ReportConfig, which means it takes too long to report an event (like Event A3 for handover). Alternatively, a high signalling load on the network may result in delays in processing the measurement reports from the UE. Internal UE processing limits may also be an issue, particularly in older devices with slower chipsets. Such holdups may cause the UE to remain connected to a weakening cell for an extended period of time, which could result in a radio link failure (RLF) before a handover can take place, increasing the risk of call dropouts or poor network performance. Because the UE doesn't move to a less loaded cell in time, delayed reporting can potentially lead to network congestion in crowded urban areas.

c. **Insufficient Measurement Reporting:** Insufficient measurement reporting can result in poor network performance and a worse user experience when the UE fails to report frequently enough or fails to notice important measurement events. Insufficient measurement setups may be the source of this, such as imposing an excessively severe Event A3 offset or an exceedingly high threshold for reporting events, which stops the UE from reporting a better neighbour cell. Insufficient measurement gaps for inter-frequency or inter-RAT measurements, particularly if the UE is extensively involved in data transmission, could be another factor. Additionally, excessive smoothing may obscure actual changes in signal circumstances, postponing or blocking event triggering, if the filterCoefficient (filterCoefficientRSRP, for example) is set excessively high (fc8). "fc8" is a value that falls within a parameter known as the measurement frequency (fc) in the context of LTE (Long-Term Evolution) mobile networks. This parameter regulates how often cell readings are reevaluated and sets off events such as handover, which occurs when a device switches between cell towers. For devices that move slowly or are primarily fixed, a larger number, of fc8, indicates a lower measurement frequency. The UE may stay on a less-than-ideal cell due to missing handover opportunities brought on by inadequate reporting, which could lead to decreased data throughput, higher latency, or even connection failures. From the standpoint of the user, this could show up as sluggish internet or dropped calls.

d. **Excessive Reporting:** Overwhelming network congestion and resource depletion might result from excessive measurement reporting. This problem frequently results from improper measurement configurations, such as setting an extremely low time-to-trigger for event-based reporting (e,g., Event A3 triggers too easily) or a noticeably short report interval for periodic reporting. Frequent measurement reporting triggers brought on by unstable signal circumstances, particularly when filtering is turned off (filterCoefficientRSRP = fc0), may also be a contributing factor. For instance, the UE may send measurement reports excessively frequently in an urban location with many small cells due to rapid signal fluctuations. Other important signalling messages (such handover commands) may be delayed because of this excessive signalling, which also uses uplink resources and puts more strain on the eNB's RRC layer. Users may experience slower network response times or even unsuccessful handovers because of the network being overloaded.

e. **Interference Issues:** The accuracy of measurement reporting can be severely impacted by interference from other networks or devices, which lowers network performance. For instance, the UE may report false readings due to co-channel interference from nearby cells utilising the same frequency, which can skew the UE's RSRP or RSRQ measurements. Similarly, if the UE is set up to measure inter-RAT frequencies, inter-RAT interference (from a neighbouring Wi-Fi network or a UMTS cell, for example) can reduce measurement accuracy. Elevated levels of interference may cause the UE to repeatedly generate events like Event A3 in crowded deployments, including urban small-cell settings, which could result in needless handovers or ping-pong effects between cells. Furthermore, the UE's measurements may continuously overestimate or underestimate signal quality if the network does not appropriately configure cell-specific offsets or interference mitigation strategies (for example, through appropriate frequency planning), which could result in poor mobility choices and a worse user experience.

f. **Handover Issues:** Inaccurate measurement reports may result in handover failures, which have a direct effect on network performance and user experience. For example, the eNB may start a handover to a cell that is not the best option if the UE reports delayed or erroneous measurements. This could lead to a handover failure or a fast return to the original cell (ping-pong effect). Another problem might be inadequate reporting, in which the UE remains on a weakening cell until a radio link failure happens because it does not promptly report a better neighbouring cell. By overwhelming the network and postponing the handover order, excessive reporting might potentially cause handover problems. Failed handovers can also result from the UE's inability to monitor target cells efficiently due to incorrectly configured measurement gaps for inter-frequency or inter-RAT handovers. These problems, particularly in high-mobility situations, can lead to missed conversations, interrupted data sessions, and poor network reliability.

g. **Congestion Management Issues:** Since measurement reports assist the eNB in distributing load among cells by starting handovers to less crowded cells, they are a crucial tool for managing congestion in LTE Heterogeneous networks. On the other hand, inadequate congestion management brought on by erroneous or delayed reporting can impair network performance. For instance, the eNB may not recognise that the serving cell is overloaded and may neglect to transfer the UE to a less crowded neighbour if the UE provides inaccurate or out-of-date RSRP values because of delayed reporting or a lack of filtering. By using up signalling resources, excessive reporting can further increase congestion by reducing the ability of other UEs to report their measurements or receive handover commands. Inadequate reporting may also make it impossible for the eNB to identify a cell that is overloaded, so losing out on opportunities to unload traffic. Especially in crowded urban settings with high user density, these problems might result in lower data speeds, higher latency, and a greater chance of connection drops.

h. **Battery Drain on the UE:** The UE may use excessive power as a result of frequent or incorrectly set measurement reporting, which will accelerate battery drain. For instance, the UE's radio must stay active for longer periods of time, using more power, if the eNB sets the UE to do frequent inter-frequency or inter-RAT measurements without adequate measurement gaps. The UE is also forced to send measurement reports more frequently when there is excessive reporting (for example, because of a short report interval or low time-to-trigger), which raises uplink signalling and power consumption. When the s-Measure parameter is set too low, the UE may have to measure nearby cells even when the signal from the serving cell is strong, which would further deplete the battery. This problem affects the user experience by necessitating more frequent charging and can be especially troublesome in places with weak signal strength, where the UE already uses more energy to stay connected.

i. **Improper Measurement Gap Configuration:** To measure inter-frequency or inter-RAT cells, the UE can halt connection with the serving cell during measurement gaps. Incorrect configuration of these gaps by the eNB may result in problems with measurement. For instance, the UE may not have enough time to precisely measure target cells if the measurement gap duration is too short (6 ms within 40 ms), which could result in results that are either wrong or incomplete. On the other hand, excessively long or frequent gaps may disrupt the UE's data transfer, lower throughput and raising latency. Delays in handovers or radio link failures may result from improper gap setup, which also makes it impossible for the UE to identify a better cell in time. Inter-frequency measurements are crucial for mobility in heterogeneous networks, which consist of a mixture of small and macro cells that operate on several frequencies.

2.8.2.3 Adjustment of The Inter-Rat HO for Traffic Steering

Inter-RAT (Radio Access Technology) handover problems which arise when a mobile device shifts between multiple types of networks (such as 2G, 3G, or 4G), can be resolved by optimising handover parameters and analysing network performance. This often includes modifying handover thresholds, neighbour cell lists, and even applying self-optimizing network (SON) algorithms.

The first problem with adjusting HO parameters is that they are normally modified to avoid low signal levels or quality to keep the connection active. However, in the context of HetNets, there may be numerous more reasons to initiate a HO, given that the coverage areas of the network layers overlap. Following the 3GPP standard, it is possible to establish different thresholds for the same HO triggering event based on specific policies. In this regard, radio-driven HOs are performed when the UE suffers a low signal strength or quality, whereas TS-driven HOs are initiated for traffic steering (TS) reasons, such as maximising user throughput. In the first situation, UE devices are typically passed over to cells of higher quality.

Table 2.3 Event configuration (3gpp 36.133)

Event	Parameter	Range		Value	
A3, A6, C2	Offset:	-30	30	-15dB	+15dB
A1, A2, A4, A5, B1	RSRP threshold	0	97	-140dBm	-44dBm
	RSRQ threshold	0	34	-19.5dB	-3dB
C1	CSI RSRP	0	97	-140dBm	-44dBm
B1, B2	UTRA RSCP	-5	91	-120dBm	-25dBm
	UTRA EcN0	0	49	-24dB	0dB
	GERAN	0	63		

In the second situation, UE devices are typically handed over even when signal conditions are favourable. Because these two instances are complementary, the operator could design two distinct sets of thresholds for a single HO triggering event in the network. Inter-Radio Access Technology (RAT) handovers (HOs) may include four thresholds: A1, A2, B1, and B2. These thresholds are used to initiate measurements and make subsequent handover decisions based on signal strength and other parameters.

2.8.3 Traffic Steering and Mobility Management with Femto Cells

It is crucial to remember that while both femtocells and picocells are small cell types used to increase the capacity and coverage of cellular networks, their deployment circumstances and certain technological features differ. While picocells are more adaptable and may be used both indoors and outdoors, especially in places with high user density or where macro cell coverage is poor, femtocells are primarily for indoor use and are frequently installed by people or small businesses to increase coverage within a house or office.

Because femto cells can be installed in enormous numbers and are often beyond an operator's control, a user could move or shut off a femto base station at any time they provide unique issues. The network must therefore adjust to a constantly shifting environment. The number of potential Physical Cell Identifiers (PCIs) may become a limitation for a high density of femto cells, which could result in unclear cell identification. In LTE (Long Term Evolution) networks, the Physical Cell Identity (PCI) runs from 0 to 503, resulting in a total of 504 unique identifications. This PCI is a 9-bit value used to identify specific cells within a base station (eNodeB). The PCI allows user equipment (UE) to locate and connect to the proper cell for communication, as well as simplify cell handovers.

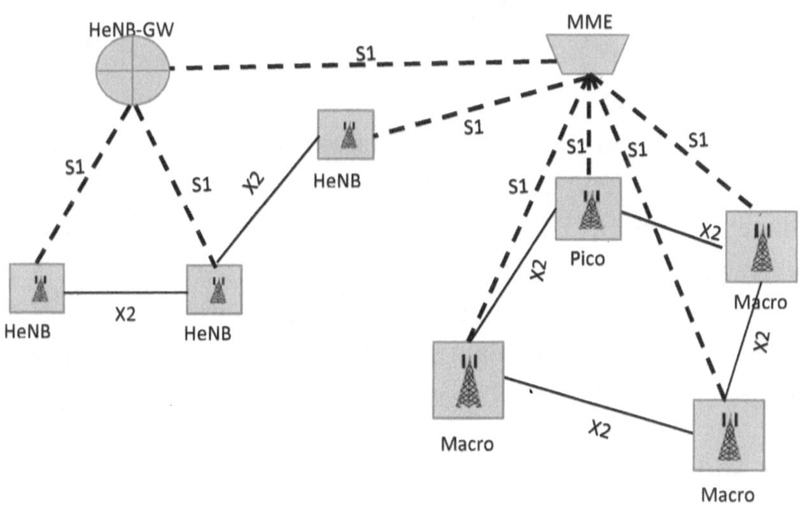

Figure 2.21 Interfaces between different LTE base station nodes used for mobility.

Moreover, as mentioned earlier, femto cells could be limited to a Closed Subscriber Group (CSG), which restricts the options for cell reselection and handover. Handling CSG femto cells is made easier by base stations' ability to broadcast PCI ranges associated with CSGs to terminals in LTE and LTE-Advanced. Moreover, UEs employ a separate search algorithm from the RAT they are camped on. The autonomous search algorithm decides when and where to seek for the allowed CSG cells. Terminals may broadcast their proximity indications to their macro base station to alert them to the presence of a femto cell they may have camped on.

To illustrate, based on past CSG proximity data using the present carrier activity fingerprint, a terminal may initiate such an indication. However, terminal providers are free to choose the precise algorithm. By selecting this option, terminals in connected mode may be able to do fewer inter-frequency measurements overall. Additionally, X2-based handovers between Home eNodeBs (HeNB) are a feature of Release 10 that improves mobility. Home eNodeB corporate installations are the primary target of this, where Home eNodeBs are deployed using the same CSG ID or configured in hybrid access mode. In general, X2-based handovers between Home eNodeBs are faster than S1-based mobility processes and reduce signalling to the core network. X2 based handover between Home eNodeBs and macros, as well as X2 connectivity through the Home eNodeB gateway (Home eNodeB-GW), are not supported in Release 10.

2.8.3.1 Mobility management in LTE Femtocell network

An eNB and a HeNB differ primarily in their coverage region and intended deployment. In an LTE network, an eNB (evolving NodeB) is a typical base station that offers extensive coverage. A femtocell, sometimes referred to as a HeNB (house eNodeB), is a low-power base station intended for more confined, localised coverage, usually in a house or small office. The growing need for high-speed connectivity, the emergence of smart buildings and IoT devices, and the continuous digital transformation in many industries are expected to fuel significant development in global indoor data traffic in the upcoming years.

The demand for better interior coverage and rising mobile data consumption are the main drivers of the significant rise in femtocell indoor data traffic. According to a GSMA research, most mobile data traffic 70% comes from indoor sources and is sent over mid-band spectrum. To provide high-performance coverage, 5G mostly relies on the 3.5 GHz frequency spectrum for indoor traffic. The 6 GHz frequency may also be used as an extra capacity layer in cities and offer coverage that is comparable to 3.5 GHz indoors, according to the paper. Femtocells are a great technique to reduce the traffic burden of macrocells.

Thus, a vast quantity of mobile terminals are utilised in indoor settings. Femtocells are a great technique to reduce the traffic burden of macrocells. However, it is difficult to control femtocell mobility when so many terminals are moving erratically, entering cells, and then departing. The effectiveness of managing the terminal's mobility has a direct impact on wireless performance. Mobility control in LTE A femtocell network is a collection of tasks for managing and keeping surveillance on mobile user terminals, or equipment (UE), in a wireless network to locate them for service delivery and to keep their connections while they are in motion [15]. Mobility management encompasses several domains, including admission control, location management, power management, Handover management, and quality of service (QoS). Due to its direct influence on battery consumption, network performance, and user experience, it is one of the most important aspects in wireless communications [16]. Location management and Handover management are two of mobility management's main pillars [7]. However, the deployment of a high number of small cells may result in increased mobility management complexity, possibly due to frequent UE handover when UEs move across the small cells' narrow coverage areas. In HetNets, where macro, micro, pico, and femto cells have varied coverage, using the same set of handover parameters for all cells and/or all UEs, as in macro-only networks, may compromise mobility performance.

For instance, when and where the handover process is started in each cell will depend on the CRE's small cell bias. As a result, HetNets need to optimise the handover parameters for each cell. Handover failure owing to reduced SINR can also result from high mobility macro cell UEs running deep inside the coverage regions of small cells before the TTT optimised for macro cells ends. High mobility macro cell UEs may not require handover (such as ping-pong) if they swiftly traverse small cell coverage zones. These facts further highlight the necessity of optimising the handover parameters according to the particular mobility state of the UE (e.g., mobility speed, crossing from macro to small cell regions).

Consequently, the trade-off between ping-pongs and handover failures is a critical problem for optimising mobility performance in HetNets; lowering handover failures would raise ping-pongs, and vice versa. Cell-specific and UE-specific handover parameter optimisation is required to manage this trade-off. When UEs are moving quickly and there are many small cells underneath the macro cells, optimisation becomes more difficult. As a result, it's critical to adjust handover parameters (such TTT) considering mobility status data. Before a handover procedure is started, a certain handover condition must hold true for a certain amount of time. The TTT parameter is essential for balancing handover performance since it avoids delays that might result in poor service quality or radio connection failures, as well as excessive handovers (ping-pong effect) when signal conditions fluctuate quickly.

To determine if a UE is in the low, medium, or high mobility state in a homogeneous macro cell network, the number of handovers within a certain time window can be compared with two distinct criteria. Because of the different cell sizes in HetNets, this is no longer feasible. For instance, estimates of mobility states are skewed towards medium and high mobility levels due to the larger densities of pico cells. Within the 3GPP, many approaches to enhance the estimation of mobility status information are still being researched. In addition to statistics of handover failures and/or ping-pongs, HetNet mobility control and management necessitates thorough monitoring and assessment of KPIs pertaining to handover parameters.

2.8.3.2 Mobility Location management

Location management includes two componenets, namely registration and paging. Registration is the task of knowing where the UE is located to manage incoming and outgoing calls. The registration process is performed via a database called the Home Locations Register (HLR). The HLR contains information about the UE and its capabilities. It also describes the home area of the user. The Visitor position Register (VLR) is an additional database that is connected to the HLR to keep track of the user's present position. Every time the user switches between areas, the VLR is updated. When the UE is setting up a call, the network might page it thanks to a procedure called paging. Before making a call or sending a message, the paging process involves sending a message to the serving BS to acquire a response from the UE. In an LTE network, the Home Subscriber Server (HSS) serves as a central database for managing subscriber information, authentication, authorisation, and service profiles. It serves as the principal store for subscriber data, allowing users to connect to the network, maintain security, and access their subscription services with ease. The HSS also enables mobility management, which allows users to move across networks.

The Home Location Register (HLR) has mostly been superseded by the Home Subscriber Server (HSS) in cellular networks, especially those that use LTE and later. The HSS provides improved functionality and scalability, which are essential for managing the complexity of emerging technologies, even though both control subscriber data. Like the rest of the LTE network, femtocells, which are small, low-power base stations, usually use the HSS for authentication and subscriber management. The Mobility Management Entity (MME) is mostly responsible for the Visitor Location Register (VLR) functionality in LTE (Long-Term Evolution) networks. LTE uses a different architecture, with the MME managing the tasks of tracking subscriber position and controlling mobility, whereas the VLR is a crucial part of 2G/3G circuit-switched networks for managing roaming users.

2.8.3.3 Femtocell Mobile modes

Idle mode: When there is no continuous service (voice, data) on the UE, this mode is in effect. After powering on and setting up its position, the UE spends most of its time in this mode, where it watches for page messages from the networks. The user interface does a reselection of BS whilst in idle mode. Cell reselection tries to maintain camping on the best available cell during the idle mode and aids in transfer registration (VLR)/MME. Following the reselection criterion, UE constantly looks for a better cell. Selection is another process of mobility. The process of choosing a suitable cell to camp on when a UE is turned on or after losing coverage is known as cell selection.

Connected mode: The process by which a femtocell, a tiny, low-power cellular base station, links to the core network of a mobile network operator (MNO) is known as "femtocell connected mode." Femtocells improve indoor coverage by setting up a dedicated link through the user's cable or DSL broadband internet. By doing this, mobile devices can connect to the femtocell network rather than the macrocell (outside) network, enhancing data and voice service in the covered region. After the UE releases its active session will switch to the idle mode. When the UE is moving with connected mode, Handover occurs from one BS to the next.

3.1.3.4 Handover management

With the growing use of femtocells to increase coverage and capacity in particular regions, LTE femtocell handover management is essential for improving network capacity in mobile communications. Femtocells can increase indoor coverage and capacity because they are short-range, low-power base stations. Their limited-service area, however, results in frequent handovers, which can degrade user experience and network performance. To lessen these difficulties, effective handover management is required.

Handover (HO) management is the main function by which wireless network support mobility and to maintain quality of service. HO enables the network to keep the UE's connection while it moves from the coverage area of one cell/sector to another. HO is the process of transferring an ongoing voice call or data session from one cell connected to the CN to another. HO is divided into two broad categories: Hard and Soft HOs. In Hard HO, current resources are released before new resources are used. However, in Soft HO, both existing and new resources are used during the HO process. It is important to note that Long Term Evolution (LTE) protocol does not use soft handover. Rather, it uses hard handover, in which a connection to a new cell is made before the connection to the old cell is disconnected. This is so that soft handover is not required because LTE's architecture is made for faster cell switching.

2.8.4 Wi-Fi Offloading

History of network technology and standardisation both indicate that not every feature of a standard will be implemented in actual networks. Suppliers and service providers will not adopt them until there are good business reasons to do so. Local Wi-Fi traffic breakout from their secure Wi-Fi networks (802.1x) has been the primary concern of most mobile providers. It is reasonable to assume that every laptop and smartphone has WLAN (WiFi) connectivity. When a Wi-Fi network connection is available, terminal data flow may be routed to Wi-Fi. end-user performance can be enhanced, and operator-licensed spectrum congestion can be decreased by offloading traffic from mobile networks. when the user or the terminal chooses a Wi-Fi connection, Wi-Fi offloading can occur without any network control. however, there are a number of advantages to adopting designated network-controlled Wi-Fi offloading as well.

Operators have chosen to employ secure SIM-based authentication, in conjunction with policy management from the Mobile Core, since there is no business or practical justification for backhauling Wi-Fi traffic to the Mobile Core. A substantial portion of what is feasible and achieved is also under the hands of device manufacturers. The decision by device manufacturers to deploy the IPsec client required for untrusted Wi-Fi connectivity took a decade. It just took that long, in their opinion, for a compelling business case to emerge. This was made possible via Wi-Fi Calling, which benefited both them and their clients.

In the 5G era, operators must embrace Wi-Fi more than ever before, as this study has already covered. The second is the arrival of a new type of carrier-grade Wi-Fi, known as Wi-Fi 9. Wi-Fi 9, also known as IEEE 802.11be or Extremely High Throughput (EHT), is the next generation of Wi-Fi technology that is currently being developed. It promises considerable gains in speed, capacity, and efficiency over Wi-Fi 6 (802.11ax). It is projected to become commercially available in the next years. Furthermore, operators will finally have a valid justification to backhaul traffic to the Mobile Core thanks to the new Access Traffic Steering, Switching & Splitting (ATSSS) 3GPPP standard.

Wi-Fi networks are underutilised because many users are not familiar with the technology and find it too difficult and complicated to make all the necessary choices to connect over a Wi-Fi network.

Table 2.4 Gateway functions for 3G,4G and 5G

Gateway Function	3G	4G	5G
Wireless Access Gateway	WAG	TWAG	TNGF
IPSec Termination Gateway	TTG	ePDG	N3IWF
Packet Gateway	GGSN	P-GW	UPF

When there are several WiFi networks in public places, the user could find it challenging to choose the best one. The network-controlled solution simplifies the process of finding and choosing a Wi-Fi network based on operator and end-user preferences. A home network, operator hot spot, or partner Wi-Fi network might all be considered the target Wi-Fi. The needs of the application in differing markets may also influence the decision to switch between Wi-Fi and cellular. It is possible to incorporate Wi-Fi into the mobile core network. The integration enables Wi-Fi subscribers to use the current service infrastructure, such as policy control and charging. When switching between Wi-Fi and cellular, the integration also aids in maintaining service continuity. The offloading scenario with integration to the mobile core network is shown in Figure 2.22

This means that when a user opens an application such as a browser, email, or Internet radio, the device chooses, if one is available, an automatically configured Wi-Fi network rather than a cellular network. As a result, Wi-Fi offloading systems give an operator the guarantee that, when possible, traffic is offloaded to preferred Wi-Fi networks without depending on user skill. An algorithm known as the Access Network Discovery and Selection Function (ANDSF) is used for the automatic selection. ANDSF is designed to help the terminal find non-3GPP access networks, such as Wi-Fi, which are usable for communication. It also gives the terminal rules that regulate connections to these networks.

2.8.4.1 Trusted and Untrusted Integration to 5G Core

Using various capabilities, trusted and untrusted WLANs interface with the 5G Core (5GC) in a Hetnet architecture to offer seamless connection for devices that can connect to Wi-Fi. The Non-3GPP Interworking Function (N3IWF) is used by untrusted access to create secure IPsec tunnels for data and control plane traffic between the UE and 5GC. On the other hand, trusted access promotes 5G connectivity through the employment of a Trusted Non-3GPP Gateway Function (TNGF) or Trusted WLAN Interworking Function (TWIF), which permits a more tightly connected integration.The simplified graphic shows Wi-Fi service integration with the new service-based 5G Core (5GC), which was introduced in 3GPP releases 15 (untrusted) and 16 (trusted). First, this design is radio network (RAN) agnostic (N1-N3) because cellular and Wi-Fi connections share the same interfaces.

The 5G core network (5GC) offers flexibility and enhanced connection through the integration of both certified and untrusted Wi-Fi (WLAN). Untrusted WLAN access employs a N3IWF (Non-3GPP Interworking Function), whereas trustworthy WLAN access connects to the 5GC using a TNGF (trustworthy Non-3GPP Gateway Function) or TWIF (Trusted WLAN Interworking Function). This integration makes it possible to seamlessly switch between cellular and Wi-Fi networks using the same N1-N3 interfaces as 3GPP access.

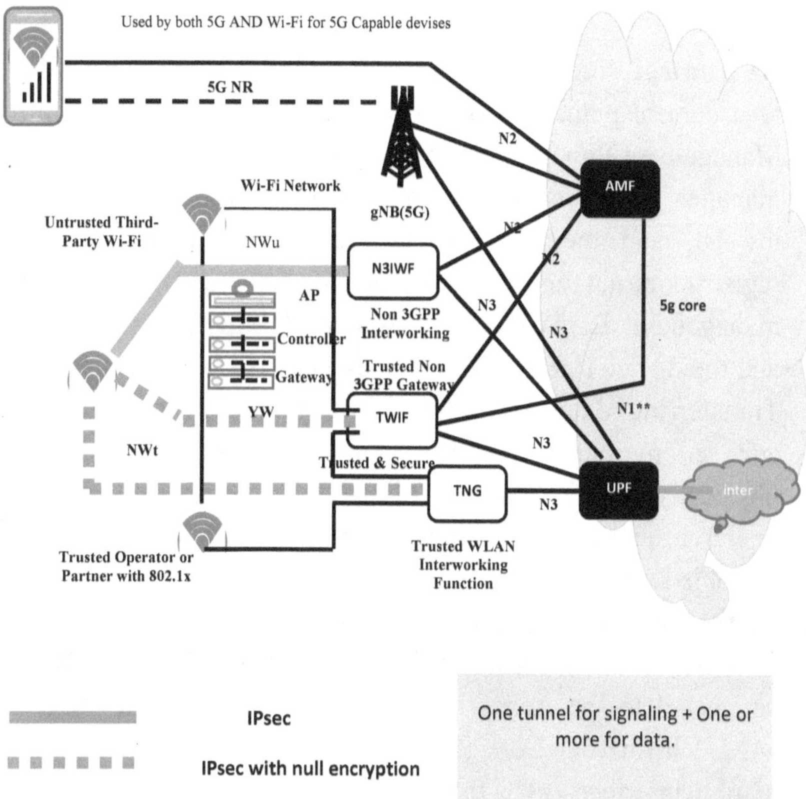

N** Replaces UE N1 for non 5G Capable devices.

Figure 2.22 Architecture Trusted and Untrusted Integration to 5G Core. 3GPP release 15 (untrusted) and 16 (trusted)

2.8.4.2 Interfaces for Wi-Fi Offloads

To reduce congestion, enhance user experience and capacity, Wi-Fi offloading in 3G, 4G, and 5G networks entails smoothly moving data traffic from cellular networks to Wi-Fi networks. Devices can transition between cellular and Wi-Fi connections depending on a number of interfaces and mechanisms, including operator policies, network conditions, and signal strength. This section describes the NI, N2, and N3 interfaces:

1. **N1 Control Plane Interface:** In 5G, the N1 interface functions as a control plane interface between the Access and Mobility Management Function (AMF) and the User Equipment (UE). It manages Non-Access Stratum (NAS) signalling, which allows the UE and the 5G core network to communicate about registration, authentication, session management, and mobility management. Between the Access and Mobility Function (AMF) and the device (UE), the N1 serves as a control plane interface. Transferring data from the UE to the AMF regarding the connection, mobility, and session is its main function. It is physically delivered to the AMF in the same manner as the N2 interface, and it is utilised for both cellular and Wi-Fi (for 5G Capable Devices).
2. **N2 Control Plane Interface:** In 5G networks, the Radio Access Network (RAN), more especially the gNodeB (gNB), is connected to the Access and Mobility Management Function (AMF) in the 5G Core (5GC) via the N2 interface, a control plane interface. It manages signalling for connection management, PDU session management, UE mobility, and UE (User Equipment) context. The N2 is the control plane interface between the access network and the 5G Core. It is primarily used for connection management, UE context and Protocol Data Unit (PDU) session management, and UE mobility management.
3. **N3 Data Plane Interface:** A 5G network's N3 interface is a user data plane interface that makes it easier for user data packets to be transmitted between the User Plane Function (UPF) and the Radio Access Network (RAN), most especially the gNodeB. It provides the real conduit for data transmission and is essential

for creating and sustaining data sessions for user traffic. The User Plane Function (UPF) in the 5G Core and the access network are connected via the N3, which is the data plane interface. The packet gateway that sends data to the internet is called the UPF.

2.8.4.3 Interworking for Wi-Fi Download

Cellular networks connect the base station (gNB) to the AMF and UPF via the N2 and N3 interfaces. Their method of connecting to the AMF and UPF for Wi-Fi is to employ the non-3GPP interworking and gateway functions (N3IWF, TNGF, and TWIF). A novel concept for non-3GPP access is brought about by 5G: multiple non-access stratum (NAS) connections across the N1 interface enable simultaneous Wi-Fi and cellular connections. The same authentication protocols, EAP-AKA' and 5G-AKA, are utilised for both Wi-Fi and cellular networks, and this is a need for the new ATSSS standard.

To support Network Access Server (NAS) messages over Wi-Fi, 5G has introduced a new EAP (Extensible Authentication Protocol) technique called EAP-5G. In order to do this, an IPsec tunnel must be established between the device and the interworking and gateway functions during device registration using IKEv2 and EAP-5G. IPsec, even with null encryption, is an innovative technique to simplify device and gateway implementations, enable mobility, and support ATSSS (Access Traffic Steering, Switching, and Splitting) in trustworthy Wi-Fi networks. Often used in conjunction with IPsec, IKEv2 (Internet Key Exchange version 2) is a VPN protocol that creates safe, encrypted connections between devices. Popular among mobile users, it is renowned for its speed, stability, and capacity to instantly re-establish connections when interrupted.

An IPsec tunnel is created using the IKEv2 and EAP-5G protocols to facilitate communication between the device and the interworking and gateway services during the registration process. NAS communications are then encapsulated over the IKEv2 protocol using the EAP-5G protocol. ATSSS and mobility are further enhanced by the advantages of a fixed anchor point in the Mobile Core. The new features for non-3GPP access will now be discussed. Once more, be aware that these features differ from those of actual gates. All these Interworking Functions may potentially be housed under a single gateway. Additionally, one vendor may supply the data plane (N3), while another may supply the control plane (N1–N2).

1. **Non-3GPP Interworking Function (N3IWF):** The Non-3GPP Interworking Function (N3IWF) in a 5G network acts as a bridge, enabling seamless connectivity between untrusted non-3GPP access networks (like Wi-Fi) and the 5G core network (5GC). It allows devices to access 5G services through non-3GPP technologies while maintaining security and proper network functionality. Like the ePDG for integration with the 4G Core, the Non-3GPP Interworking Function (N3IWF) is the IPsec tunnel termination node for 5G. It is situated within the Mobile Core and uses the N1 and N2 interfaces to communicate with the Access and Mobility Function (AMF) control plane. It uses the N3 interface to communicate with the User Plane Function (UPF) for the data plane.

2. **Trusted Non-3GPP Gateway Function (TNGF):** The 5G equivalent of the Wireless Access Gateway (WAG), which provides trusted access to the 4G Core, is the trusted non-3GPP Gateway Function (TNGF). The TNGF connects to the Access and Mobility Function (AMF) control plane via the N1 and N2 interface and is situated in a trusted environment, which is frequently the Wi-Fi network. It uses the N3 interface to communicate with the User Plane Function (UPF) for the data

plane. As previously mentioned, an IPsec tunnel with null encryption is used to connect the device and the TGNF.

3. **Trusted WLAN Interworking Function (TWIF):** A 5G network feature called the Trusted WLAN Interworking Function (TWIF) enables legacy devices to connect to the 5G core network even though they might not be able to support 5G NAS signalling over trusted Wi-Fi. It serves as a bridge, allowing these N5CW (non-5G-capable over WLAN) devices to access 5G services via a reliable Wi-Fi connection. This feature for legacy device compatibility is the trustworthy WLAN Interworking Function (TWIF). This is to address the possibility that some devices could be able to authenticate 5G SIMs but not 5G NAS signalling via reliable Wi-Fi. The EAP-5G and IKEv2 protocols are not supported by these devices. Such devices are referred to as non-5G-Capable over WLAN (N5CW) by 3GPP. On behalf of these kinds of devices, the TWIF communicates with the AMF via NAS messages and houses the NAS protocol stack. The TWIF connects to the Access and Mobility Function (AMF) control plane via the N1 and N2 interface and is situated in a trusted environment, which is often is the Wi-Fi network.

The data plane uses the N3 interface to communicate with the User Plane Function (UPF). An IPsec tunnel with NULL encryption is used to connect the device to the TWIF, same as in the case of TNGF. When choosing a roaming network while travelling overseas, the options for cellular and Wi-Fi networks are comparable. Based on the configuration supplied by the home operator, the roaming network is automatically chosen. The user still can force the mobile device to switch to a different roaming network and veto the automated solution. Similarly, when it comes to choosing Wi-Fi networks, most users will be satisfied with the automatic pick, but they can still force the device to utilise a certain access network.

The operator's current network architecture can offer authentication, authorisation, and accounting (AAA). The Subscriber Identity Module (SIM) card may be used for authentication, which is transparent to end users. Numerous authentication methods are currently in use, such as SIM and certificate-based authentication, captive web portal-based login sites that require a username and password, and others. Many consumers continue to use cellular networks rather than Wi-Fi offloading if they must manually enter their account and password to access services via a Wi-Fi network. Therefore, offloading traffic to a Wi-Fi solution is made possible by smooth user authentication, which also enhances usability.

2.9 Multi-tier Interference Mitigation

The management of interference in heterogeneous networks (HetNet's), which integrate many cell types (such as macro, pico, and femto) using the same frequency, is the main goal of LTE (Long-Term Evolution) multi-tier interference mitigation. To reduce interference and enhance network performance, strategies include resource partitioning, power control, and eICIC. The fundamental elements that allow HetNet's to accomplish the promised capacity and coverage gains are the multi-tier structure with a distributed hierarchy of infrastructure devices and the efficient use of several radio access technologies (RATs), as discussed in the preceding sections. This section concentrates on the problem of effective spectrum reuse across different single-RAT system levels. Utilising the available spectrum across several RATs may boost network capacity, but optimising HetNet capacity also necessitates judiciously reusing scarce spectral resources throughout the multiple layers of a single RAT system.

2.9.1 Multi-tier Spectral Reuse Scenarios

By segmenting a cell into sections and allocating distinct frequency reuse factors to each, multi-tier spectral reuse solutions in LTE (Long Term Evolution) networks seek to maximise spectral efficiency and control interference. This strategy helps reduce Inter-Cell Interference (ICI) and enhance performance for cell-edge users, especially when Fractional Frequency Reuse (FFR) is utilised. Different spectral reuse patterns may be present in a multi-tiered network. For example, devices operating on separate tiers can either utilise distinct frequencies and function with "no reuse" or they can implement "full spectral reuse" and use the same frequency across all tiers. The biggest capacity boost can be achieved by avoiding the severe cross-tier interference that results from sharing the same spectrum by use of orthogonal frequencies across tiers.

When spectrum-rich operators have numerous frequency carriers, the best way to arrange their spectrum is to assign distinct operating frequencies to first-tier devices (such as MBSs) and second-tier devices (such as FAPs). By doing this, cross-tier interference is avoided. Second-tier devices, such as PBSs and FAPs, can choose from a variety of frequencies if there is more spectrum available, in addition to avoiding interference from same-tier devices. Spectrum limitations, however, prevent most network operators from affording to develop distinct carriers across tiers. Smart single frequency multi-tier deployments, where a single carrier is used sparingly across tiers, are therefore essential to the success of multi-tier systems.

2.9.2 Cross-tier Interference

When all first-tier MBSs and second-tier FAPs, PBSs, or RSs use the same spectrum simultaneously, co-channel interference can seriously affect network performance. Due to the additional coverage gaps that FAPs create for users without access authorisation, inter-tier interference is much more conspicuous in closed subscriber group (CSG) FAP installations. As the FAP deployment density increases, a greater proportion of macro-users are driven into outage due to the increased interference, which leads to unsatisfactory macro performance.

While FAPs provide greater indoor coverage and a capacity gain, we find that a significant portion of outdoor users are forced into outage due to co-channel interference from FAP transmissions when comparing the performance of deployments with the femtocell overlay versus those with macro-only coverage. Therefore, interference reduction techniques are essential to enable cochannel deployments at a scale where a suitable density of FAPs may be supported.

2.9.3 Network Synchronization for IM

The synchronisation of network devices is necessary for most interference management methods to work without degrading performance. In OFDM (Orthogonal Frequency-Division Multiplexing) systems, the cyclic prefix (CP) is an essential component that guarantees the orthogonality of subcarriers and helps reduce the impacts of multipath propagation. The CP is a duplicate of the OFDM symbol's end section, which is appended to the symbol's beginning. To prevent interference between neighbouring symbols, this prefix serves as a guard band.

The idea behind employing a cyclic prefix with OFDM is simple, but it necessitates more processing complexity in both the transmitter and the receiver. To achieve the Cyclic Prefix, the final few OFDM symbol samples are placed both before and after the commencement. This comes between any two distinct OFDM symbols, the cyclic prefix symbols are positioned to serve as a guard interval. The impacts of the previous symbol's ISI are lessened or even eliminated with the help of this guard interval.

The resulting CP-OFDM data is periodic or cyclic, in contrast to regular OFDM, because the prefixed symbol is cyclical in nature, meaning that the CP symbol repeats the conclusion of the OFDM data. Briefly the cyclic prefix serves two primary purposes.

- The CP is frequency-selective, meaning it separates several OFDM blocks from one another when the wireless channel has many pathways.
- The channel-based linear convolution is converted to a circular convolution via the CP.

A copy of the OFDM symbol's end section comes before each OFDM symbol owing to the creation of the cyclic prefix. Different systems offer varying lengths of OFDM cyclic prefixes. There are two lengths available in LTE, for instance: standard and extended. A third, extended length was added after Release 8, although it is not often utilised.

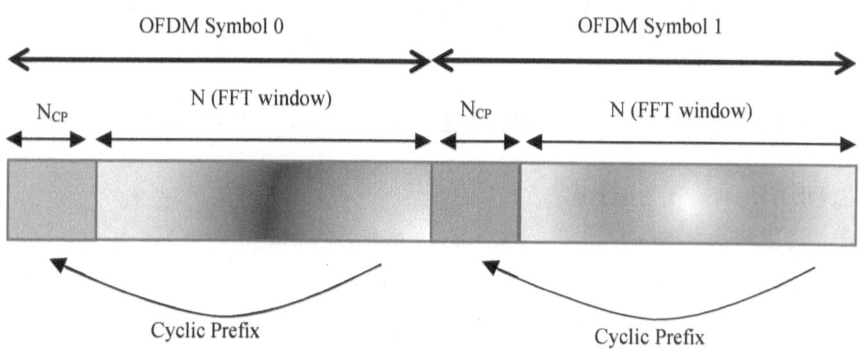

Figure 2.22 OFDM cyclic prefix

Let us examine the signal that results. As we can see, a cyclic prefix separates two adjacent symbols, and the cyclic prefix includes a duplicate of each symbol's end. Nevertheless, we are cognisant that the CP is a duplicate of the OFDM symbol's end. A circular convolution within the OFDM FFT window can therefore be understood as the linear convolution over the CP: Channel impulse response wrapping around the FFT window and blurring the end of the block into the beginning is what happens when it is above the CP area.[16] illustrates how Femtocell networks can be synchronised with the overlay macro network by leveraging existing over-the-air synchronisation technologies in the overlay system.

However, it is also noted that simply aligning the time reference of each layer with a global reference is insufficient to guarantee synchronised interference across levels. For broadcasts from devices across tiers to reach the recipient concurrently, it is essential to maintain the relative time alignment across devices in this scenario. The way that devices in a single tier network adjust their uplink transmissions based on their distance from the base station to align with the receiving base station is analogous to this need. In the same line of thought, time synchronisation between devices across tiers is also necessary to guarantee synchronous interference in multi-tier networks. Further details are available in [16].

2.10 Techniques for Mitigating Interference Overview

2.10.1 Power Control

Transmit power regulation is one of the fundamental issues facing modern wireless communications. Power control refers to the application of algorithms and techniques that control and alter transmit power coming from user equipment (UE) and base stations. Elevated levels of it could, for example, lead to a high signal-to-interference-noise ratio (SINR), which could lead to interference. A signal with a lower SINR is also likely to be badly received. To dynamically change the SINR, we use power control strategies.

The advantages include prolonging the UE's battery life and ensuring an adequate quality of service. In addition to other considerations, interference limits service quality. The successful deployment of power control strategies have the advantage of lowering interference between UEs. Other advantages include lowering co-channel interference [7], maximising cell capacity, and minimising the UE mean transmit power.

Power control is an important interference management technique since it can be accomplished without network-level synchronisation. Changing the transmission power level of a FAP allows the operator to adjust its coverage range and the amount of interference it produces in the network. Higher transmission power has the potential to improve signal quality and increase coverage, but it can also cause significant interference with neighbouring users. As a result, while maintaining femtocell performance, properly adjusting the FAP transmission power level can help manage interference from FAPs to macro-users. Three separate power control schemes are often used: 1). Fixed power level operation. 2). Femto-QoS power control, which adjusts the femto power to maintain a minimum specified data rate for the femto user; and 3). Macro-QoS power control, which controls the FAP power to guarantee the macro user's QoS is not negatively impacted. In the past, CDMA-based mobile communication systems like WCDMA and CDMA2000 have employed dynamic transmit-power regulation to account for changes in the instantaneous channel conditions. power to make up for changes and discrepancies in the current channel conditions. Maintaining a (near) constant Eb/No at the receiver is the goal of these changes to transmit data properly without an excessively high error probability. In theory, when the radio link is subject to bad radio conditions, transmit-power control raises the transmitter's power (and vice versa). As shown in Figure 2.23, the transmit power is therefore essentially inversely related to the channel quality.

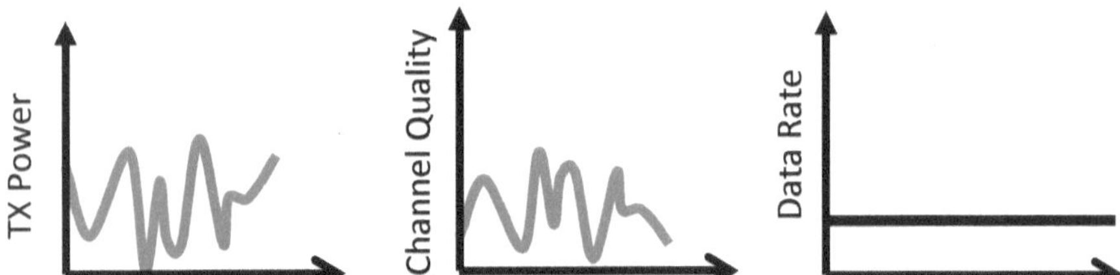

Figure 2.23 Power Control

This produces a data rate that is constant despite changes in the channel. This is a useful feature for services like circuit-switched voice. One kind of link adaptation is transmit-power control, which involves modifying transmission parameters, in this example the transmit power, to adjust to variations in the instantaneous channel conditions and keep the received Eb/No at the intended level. However, it is not always necessary to deliver a specific constant data rate across a radio link in mobile communication, particularly when it comes to packet-data traffic. Instead, the data rate offered via the radio interface should be as "high as possible" as perceived by the user. In reality, (short-term) fluctuations in the data rate are frequently not a problem, even for common "constant rate" services like speech and video, provided that the average data rate stays constant, assuming averaging over a brief period.

An alternative to transmit-power control in these situations that is, when a constant data rate is not necessary is link adaptation using dynamic rate control. The goal of rate control is not to maintain a constant instantaneous radio-link data rate under all instantaneous channel conditions. Rather, rate control dynamically modifies the data rate to account for changing channel circumstances. When the channel circumstances are favourable, the data rate is higher, and vice versa. As a result, rate control keeps the Eb/No-P/R at the target level by modifying the data rate R rather than the power P. Figure 2.24 provides an illustration of this. Rate control has been demonstrated to be more effective than power control.

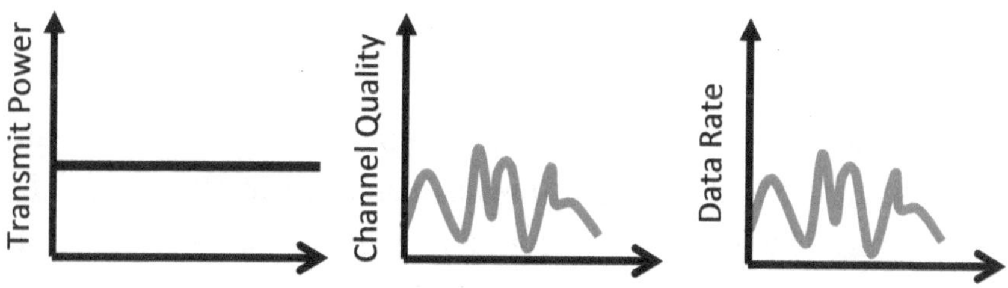

Figure 2.24 Rate Control

The power amplifier is always transmitting at full power and is thus being used efficiently, according to the principle of rate control. On the other hand, power control causes power to be below its maximum, which prevents the power amplifier from being used effectively in most circumstances. The modulation method and/or the channel coding rate are used to control the radio-link data rate in practice. Under favourable radio-link conditions, the receiver's Eb/No E is high, and the radio link's bandwidth is the primary constraint on data rate. Higher-order modulation, such as 16QAM or 64QAM, in conjunction with a high coding rate is therefore necessary in these circumstances. Similarly, QPSK and low-rate coding are employed when radio-link conditions are bad. For this reason, adaptive modulation and coding (AMC) is another name for link adaptability by rate control.

By constantly modifying modulation schemes and coding rates to correspond with shifting radio link circumstances, adaptive modulation and coding (AMC) maximises throughput and reliability while optimising data transmission in heterogeneous networks.

AMC chooses reliable, lower-throughput modulation and coding schemes (MCS) for bad conditions and effective, higher-throughput MCS for good conditions based on an accurate estimation of channel quality. This method is essential to contemporary cellular networks like 5G and LTE, and it is rapidly using more sophisticated methods, such as machine learning, to adapt more intelligently.

In a heterogeneous network (HetNet), rate control modifies the user's achievable data rate, whereas power control modifies the transmit power level. While rate control concentrates on controlling the data throughput or bandwidth to satisfy user expectations and optimise network throughput, power management is essential for overcoming channel impairments and minimising interference to guarantee signal quality. Joint rate and power control systems are employed in HetNets to optimise overall network performance, including energy efficiency and user experience, and to manage the complexity of numerous cell tiers (such as macrocells and femtocells). This entails regulating interference between femtocells and macrocells, as well as within femtocells, utilising approaches like as adaptive power regulation and distributed algorithms. The goal is to improve the user experience, increase network capacity, and reduce energy consumption.

2.10.2 Cell Association

Cellular networks are becoming more heterogeneous as more small cells are deployed into macro base stations (MBS) to satisfy rapidly increasing traffic needs. To make use of the benefits of these small cells, mobile users ought to be offloaded onto small base stations (BSs), which are often lightly populated and can provide a greater data throughput by presenting mobile users with many more channels than an MBS. Likewise, a more balanced cell association will alleviate the load on the MBS, letting it to service its remaining customers more effectively.

When FAP power control is enforced, mobile stations may be required to execute cell re-association after losing power at their supplying base stations. Users can, for example, select the maximum received signal level, following the same rules as in the original association operation. Various criteria can also be utilised to improve coverage and ensure that adequate traffic is offloaded to low-power devices, such as only performing re-association when the user is offline. Cell association and power control can connect to deliver more advanced interference mitigation. For example, a clustering optimisation process might be used to organise users based on their ability to prevent interference by effectively managing the power settings at base stations for a certain geographic distribution of users.

Heterogeneous Networks (HetNet's) are now regarded as an important architecture of fifth generation (5G) and future generation networks, having been demonstrated to be a potential paradigm for improving data rates and user capacity. In HetNet's, small cells, also known as small base stations (SBSs), are built within macro base stations (MBSs) to improve coverage in areas that are underserved.

The primary distinctions between 4G and 5G HetNet's include greater data speeds, lower latency, expanded capacity, improved coverage, and intelligent network management, among other things. 5G HetNet's enable a variety of new use cases and applications that were not possible or realistic with 4G networks. These, include eMBB (Enhanced Mobile Broadband), mMTC (Massive Machine Type Communication), and uRLLC (Ultra Reliable & Low Latency Communication). Addressing the issues of improving Quality of Service (QoS), load balance, throughput, and interference mitigation in 5G HetNet's is even more important. As a result, several research have been conducted to address the issues and fulfil the standards of 5G and future generation networks [7]. One essential method for improving 5G success is to build HetNet's, which include Base Stations (BSs) with high transmission power, such as MBS, as well as those with lower transmission power, such as PBSs and Femto cells.

The key benefits of establishing HetNet's include increased network capacity and improved link quality for users near BSs [7]. Nonetheless, traditional cellular user association methods tend to result in the bulk of users associating with MBS due to their larger transmission capacity, even when smaller cells such as pico or Femto Cells are closer. This usual association strategy causes a load imbalance, with MBSs overload and SBSs having fewer connections [8]. Additionally, the MBS's cell-edge users' throughput decreases, which often results in call dropouts and cell-edge customers experiencing outages.

Additionally, many other MBS customers experience low data rates, which lowers the average network performance. As a result, it is difficult for operators to uphold the claim of 5G standards as well as for users to receive the optimum service [9]. The load-balancing and throughput issues of 5G and future networks can be resolved using a unique load-awareness-based cell association technique that bases the association on proximity rather than the conventional RSRP and its derivatives. Therefore, improving load allocation among base stations while simultaneously guaranteeing QoS provisioning more especially, throughput for 5G and future networks is the main goal of this technique. It is important to recognise some limitations, too, such as the use of a random user deployment model to simulate mobility, which calls for greater research to produce more realistic depictions of real-world situations.

2.10.3 Low Duty Cycle

To conserve power, wireless communication devices that use Low Duty Cycle (LDC) alternate between active and sleep modes on a regular basis. LDC techniques can be used in the context of interference management to lessen the effects of interference on wireless systems, especially when coexistence is incorporated. To reduce overlap with interfering signals, this entails tactics like shortening active periods or modifying transmission timing. Even with lower FAP transmission power, growing device density in heterogeneous networks can lead to notable aggregated interference. In an urban area with multi-story buildings, for case in point, there might be approximately 800 FAPs deployed per sector within a 500 m cell. The current standards require all FAPs, active and dormant, to transmit control signalling messages concurrently with the macro-BSs, which prevents the widespread deployment of FAPs in urban areas. High FAP densities invariably lead to an unacceptable degree of interference on control channels. Control channel interference can be reduced by transmitting FAP control channels with a low duty cycle and requiring neighbouring FAPs to alternate broadcast of their control channels. Using this method, the corresponding FAP density is "virtually" reduced.

2.10.4 Advanced Time-frequency Planning

To optimise capacity and spectral efficiency, control interference, and balance network load, LTE-Advanced (LTE-A) time-frequency planning entails dynamically allocating radio resources by coordinating the time and frequency domains. Coordinated Multi-Point (CoMP) for cell-edge performance, MIMO for spatial multiplexing, and carrier aggregation for increased bandwidth are important strategies. AI-based analytics and Self-organising Networks (SON) automate resource allocation and performance optimisation depending on current traffic and interference conditions.

LTE. Allocating radio resources in both the time and frequency domains to maximise network, particularly when considering elements like relay nodes and heterogeneous networks, this covers techniques for carrier aggregation, frequency reuse, and interference management in both time and frequency dimensions. For example, many macro users may experience failures due to interference with CSG FAPs and femtocell overlay networks. Cross-tier interference can be easily avoided by designating distinct time-frequency zones for first and second tier devices, such as MBSs and FAPs, respectively. A more efficient solution would be a time-frequency allocation in which FAPs do not transmit and the resource block is thus free of FAP interference. Such a femto-free zone (FFZ) can be set up for macro users who encounter femto interference.

2.11 Status of heterogeneous networks Wireless Standards

To accelerate the utilisation of heterogeneous networks, wireless standards organisations are diligently looking for "enabling" technologies. In this regard, 3GPP has advanced significantly over the past a considerable amount of time.

Table 2.5 HetNet 3gpp Advancements

Release	3gpp Advancement
Release 10	A major improvement to 3GPP standards, LTE Release 10 (commonly referred to as LTE-Advanced) included important technology that made Heterogeneous Networks (HetNets) possible. These technologies include relay nodes, enhanced Inter-Cell Interference Coordination (ICIC), carrier aggregation, and improvements to MIMO (Multiple-Input Multiple-Output). Enhancing spectral efficiency, adding low-power nodes (such as femtocells and picocells) to boost capacity, and controlling interference in these multi-layered networks are the main goals of these HetNets characteristics.
Release 11	By adding and enhancing features like Coordinated Multi-Point (CoMP), Further Enhanced Inter-Cell Interference Coordination (FeICIC), and the Enhanced Physical Downlink Control Channel (E-PDCCH), LTE Release 11 improved support for heterogeneous networks (HetNets). HetNets are known for their ability to better manage interference in mixed cell situations, increase system spectrum efficiency, and enhance the cell-edge user experience.
Release 12	Through features like enhanced inter-cell interference coordination (eICIC), Coordinated Multipoint (CoMP) operation, and mobility state estimation enhancements to handle the increased cell changes in small cell deployments, LTE Release 12 on HetNets (Heterogeneous Networks) aims to improve user mobility and performance in dense environments. Additionally, HetNets benefit from Release 12's general LTE-enhanced advancements, such as 256-QAM and enhanced MIMO (Multiple-Input Multiple-Output) feedback enhancements, which

	increase spectral efficiency across cells of all sizes. Additionally, Release 12 addressed the scalability challenges of X2 connections between eNBs and HeNBs. It introduced the concept of X2 Gateways (X2GW) to allow an eNB to connect to multiple HeNBs through a single or a few X2GWs.
Release 13	By expanding Carrier Aggregation (CA) to 32 Component Carriers (CCs) for higher data rates in both licensed and unlicensed bands, enabling Licensed-Assisted Access (LAA) to unlicensed spectrum, and enhancing Dual Connectivity (DC) for better traffic steering and multi-vendor support, LTE Release 13 greatly improved Heterogeneous Networks (HetNets). Because of these developments, HetNets were able to more effectively integrate several cell types (macro, micro, pico, and femto), supporting the higher capacity and performance that growing mobile broadband services require.Release 13 includes specifications for managing H(e)NBs, including secure IP connections, registration procedures with the Home eNodeB Management System (HeMS), and functionalities for enabling mobile operators to offer more advanced services.
Release 14	The use of 3GPP Release 14 standard features for coordinated and improved small and macro cell deployments is known as LTE Release 14 on a heterogeneous network (HetNet). By better coordinating control information (PDCCH) and physical channels between large and small cells, this release improved radio interfaces with features like carrier aggregation with cross-carrier scheduling and eLAA (evolved licensed assisted access), which help manage a variety of cell types and spectrum in HetNets. Release 14 also addressed the establishment and management of IPsec tunnels between H(e)NBs and the Security Gateway (SeGW) for secure communication.
Release 15	By adding technologies like E-UTRAN – NR Dual Connectivity (EN-DC), Release 15 enhanced LTE capabilities and made it possible to integrate 5G New Radio (NR) with pre-existing LTE infrastructure, improving services and opening the door for future 5G services within the HetNet. By using cross-carrier scheduling to distribute the load between macro and small cells, HetNets in LTE Release 15 enhance overall performance and reduce PDCCH interference.
Release 16	The focus of LTE Release 16 (Rel-16) is on features that expand and build upon 5G New Radio (NR) capabilities rather than improvements for LTE itself. It does, however, bring some LTE improvements and technologies that complement 5G NR in HetNets, namely Dynamic Spectrum Sharing (DSS), which enables the sharing of spectrum between LTE and NR. Along with improved mobility and dual connectivity for faster data rates and system capacity, Release 16 also makes it easier for LTE to cooperate with NR in complex settings. Release 16 continues to build upon and refine the functionalities and management of H(e)NBs, with a focus on service requirements, system architecture, and various operational aspects like mobility, performance, and lawful interception.
Release 17	LTE Release 17 does not expressly mention HetNets; rather, LTE (Long-Term Evolution) is a foundational technology that can function within HetNets, which are networks made up of various access technologies (such as 4G LTE and 5G NR). Release 17 focusses on expanding 5G New Radio (NR) and IoT capabilities, but it

	also includes some LTE advancements, such as those for NB-IoT and LTE-M, which improve power saving features, data speeds, and coexistence with 5G NR. Release 17's support for NTN and RedCap devices directly impacts Heterogeneous Networks (HetNets). HetNet's, by definition, involve a mix of different technologies and coverage areas. In addition, Release 17 enhances the flexibility and reach of HetNet's by integrating satellite connectivity and supporting a wider range of device types.
Release 18	LTE update 18, also known as the first 5G-Advanced update, adds new features to 5G systems, but the concept of "LTE on HetNet" is more of a background technology that LTE and now 5G build on and develop. HetNets (Heterogeneous Networks), which mix multiple cell types such as macro cells, small cells, and Wi-Fi, have long been a target for improved coordination and efficiency. New features in Release 18 include AI/ML for network energy savings, duplex evolution, and advanced MIMO and RedCap upgrades, which will indirectly improve the performance and management of these HetNets by offering more complex tools for managing varied access networks. Release 18 maintains the support for H(e)NBs while integrating it with the broader 5G-Advanced framework and introducing new features and enhancements for the overall 5G ecosystem.
Release 19	LTE Release 19 (Rel-19) is the second release of 5G-Advanced, a continuation of the 5G system's growth, with an estimated completion date of December 2025.The first edition of this boob will update. While the query references LTE, Rel-19 is described as part of the broader 5G system, with the primary goal of introducing new features and advancements to 5G capabilities, with some overlap with future 6G requirements. Non-terrestrial networks (NTN), artificial intelligence/machine learning (AI/ML) integration, RedCap (Reduced Capability) device upgrades, and further AI/ML applications in the network are all critical topics. Release 19, the focus shifts from the traditional H(e)NB (Home NodeB/evolved NodeB) to broader concepts like Non-Terrestrial Networks (NTN) and energy saving for sustainability.
Release 20	Release 20, the term H(e)NB, which refers to Home NodeB (HNB) and Home eNodeB (HeNB), is not specifically mentioned as a major focus or feature. Heterogeneous networks (HetNets), which comprise a variety of cell types, have been implemented and studied since the launch of LTE, but Rel-20 focusses on the future rather than the past.
Release 21	Release 21, H(e)NB (Home (e)NodeB) functionality is not explicitly mentioned as a major focus or new feature. Release 21 primarily addresses advancements in 5G-Advanced, with a strong emphasis on enhancements for enhanced mobile broadband (eMBB), massive machine-type communications (mMTC), and ultra-reliable low-latency communications (URLLC). While H(e)NBs (specifically HeNBs in LTE) were important in previous releases for providing localized coverage, their role has diminished with the widespread deployment of 5G and the evolution towards network densification and small cells integrated into the broader 5G architecture.

For example, additional features have been introduced to the standard to support femtocell deployments (called H(e)NB in 3GPP terminology). Ongoing work items in 3GPP Releases.

2.11.1 Future Research Directions

Optimising energy efficiency, improving network throughput, vertical and horizontal optimization, boosting security, and adjusting to new technologies like 5G and beyond, as well as the Internet of Things (IoT), should be the main goals of future research on HetNet's (Heterogeneous Networks). Research should also investigate network slicing for customised services, multi-access edge computing (MEC) for low-latency applications, artificial intelligence (AI), and machine learning for network optimisation.

2.11.2 Vertical and Horizontal Optimization

The goal of HetNet (Heterogeneous Network) optimisation, namely vertical and horizontal optimisation, is to enhance user experience and network performance in a network with a variety of access mechanisms. While horizontal optimisation concentrates on resource allocation and management across various access technologies inside the HetNet, vertical optimisation aims to improve the performance of components or access technologies. It is expected that future heterogeneous networks will be optimised in both vertical and horizontal directions. Vertical optimisation includes extra network architecture flattening and/or cross-layer protocol optimisation.

This approach is necessary to reduce network latency and improve end-to-end QoS. Horizontal optimisation includes cooperative signal processing across many sectors or cells, which may be from the same tier or other associated tiers. For example, a super base station can serve as a central location for multi-level interference control coordination. Because the internal connection of the super base station takes the place of the X2 interface between base stations, it may be a more cost-effective way to implement CoMP and carrier aggregation in future heterogeneous networks.

2.12.3 Service on Network Edge

Upcoming technologies that future heterogeneous networks must fully utilise include virtualisation and multi-core processing systems. Virtualisation enables multi-core processors to support many operating systems simultaneously, and dynamic resource allocation for computation may adapt to the constantly shifting service requirements of the network. Reducing latency is a top priority for wireless broadband networks. We expect this to lead to further migration of services to the network edge. There are two factors that make this pattern noteworthy. Initially, it offers new sources of income for service providers, such as localised radio resource management for improved user experience, value-added data processing at the network edge, and user access to local content. With the proliferation of media-intensive apps, network congestion in the core network may be significantly reduced.

2.11.4 Device-centric System Design

A recurring issue for heterogeneous networks is the management and coordination of multiple access networks. Using inter-system interfaces, the centralised or "network centric" model employed by conventional cellular networks ensures that client devices function mostly uninterrupted across many networks. The main disadvantage is that it cannot scale effectively in a heterogeneous network composed of multiple access networks. A dispersed or "device-centric" approach is expected to be used by future heterogeneous networks. The creation of sophisticated connection managers and intelligent multi-mode devices is a key technological goal.

2.12.4 Heterogeneous Network for M2M Communications

In M2M (Machine-to-Machine) communications, a heterogeneous network (HetNet) is a network design that connects many devices for diverse Internet of Things (IoT) applications by combining multiple wireless network types and technologies. By solving the difficulties of M2M communication, this integration enables effective resource use, enhanced capacity, and better coverage. With the deployment of more Wi-Fi APs, FAPs, and PBSs, the coverage of heterogeneous networks will continue to improve, particularly inside buildings. This enables new services, such machine-to-machine (M2M) communication-based services, to connect devices in locations that were previously inaccessible to cellular networks (e.g., basements of buildings). Furthermore, in the future, we expect to see some M2M networks integrated into heterogeneous networks. This allows M2M aggregation points to serve their intended M2M function and provide "hotspot" access. Wireless data consumption expected to increase with mobile M2M connections.

2.13 Conclusion

The demand for wireless data appears to be increasing as wireless broadband is deployed globally. In addition to providing network operators with the lowest total cost of ownership, heterogeneous networks have tremendous potential for meeting customer demands. In standardisation forums, the wireless community is attempting to address the myriad technical difficulties that arise at the network and air interface layers. We expect heterogeneous networks to play a significant role in bringing inexpensive wireless broadband to every region of the world. Addressing existing and forthcoming regulatory requirements is especially important as heterogeneous networks integrate into the mainstream wireless communication infrastructure. In a diverse networking environment, the issue of legal intercept assistance will become more complex. In addition, the network's new services (such as m-Health) may increase service providers' liability. These new criteria must be considered and managed in future research. Wireless broadband data traffic is becoming the primary driver of both consumer purchases and network operator deployments, fuelled by customer demand for premium mobile internet services, as mobile users want access to streaming video and other high-quality content, just as they do with fixed broadband internet connections.

References

1. S. K. Joshi, K. B. S. Manosha, M. Codreanu, and M. Latva-aho, "Dynamic inter-operator spectrum sharing via Lyapunov optimization," IEEE Trans. Wirel. Commun., vol. 16, no. 10, pp. 6365-6381, Oct. 2017, DOI: 10.1109/TWC.2017.2722999.
2. R. K. Saha, "Interweave shared-use model for dynamic spectrum access in millimeter-wave mobile systems for 6G," in Proc. 2020 IEEE 92nd Vehicular Technology Conference (VTC2020- Fall), Victoria, BC, Canada, 18-Nov.–16 Dec. 2020, pp. 1-6, Doi: 10.1109/VTC2020-Fall49728.2020.9348671.
3. Ahmed, Z. Elsaraf, F. A. Khan, and Q. Z. Ahmed, "Cooperative non-orthogonal multiple access for beyond 5gnetworks," IEEE Open Journal of the Communications Society, vol. 2, pp. 990-999, 2021, doi:10.1109/OJCOMS.2021.3075081.
4. 4 R. K. Saha, "3D spatial reuse of multi-millimeterwave spectra by ultra-dense in-building small cells for spectral and energy efficiencies of future 6g mobile networks," Energies, vol. 13, no. 7, 2020, Art. no. 1748, doi: 10.3390/EN13071748.
5. R. K. Saha and C. Aswakul, "A novel frequency reuse technique for in-building small cells in dense heterogeneous networks,'" IEEJ Trans. Elect. Electron. Eng., vol. 13, no. 1, pp. 98111, Jan. 2018, doi: 10.1002/tee.22503.
6. W. Liang, K. D. Wang, J. Shi, L. Li, and G. K. Karagiannidis, "Distributed sequential coalition formation algorithm for spectrum allocation in underlay cognitive radio networks," IEEE Access, vol.7, pp. 56803-56816, 2019.
7. R. K. Saha, "Hybrid interweave-underlay millimeterwave spectrum access in multi-operator cognitive radio networks toward 6G," in Proc. Fifteenth International Conference on Systems and Networks Communications (ICSNC), Porto, Portugal, 18-22 Oct. 2020, pp. 42-48.

8. R. K. Saha, "A technique for massive spectrum sharing with ultra-dense in-building small cells in 5G era," in Proc. IEEE 90th Veh. Technol. Conf. (VTC-Fall), Honolulu, HI, USA, Sep. 2019.
9. R. K. Saha, "A tactic for architectural exploitation of indoor small cells for dynamic spectrum sharing in 5G," IEEE Access, vol. 8, pp. 15056-15071, January 2020, doi: 10.1109/ACCESS.2020.2966230.
10. D. -W. Yue, H. H. Nguyen, and Y. Sun, "mmWave doubly-massive-mimo communications enhanced with an intelligent reflecting surface: asymptotic analysis," IEEE Access, vol. 8, pp. 183774-183786, 2020, Doi: 10.1109/ACCESS.2020.3029244. M. R. Hassan, G. C. Karmakar, J. Kamruzzaman and
11. B. Srinivasan, "Exclusive use spectrum access trading models in cognitive radio networks: a survey," IEEE Commun. Surv. Tuts., vol. 19, no. 4, pp. 2192-2231, 2017, doi: 10.1109/COMST.2017.2725960.
12. A. Sharmila and P. Dananjayan, "Spectrum sharing techniques in cognitive radio networks – A survey," in Proc. 2019 IEEE International Conference on System, Computation, Automation and Networking (ICSCAN), Puducherry, India, 2019, pp. 1-4.
13. 3GPP, "Evolved Universal Terrestrial Radio Access (EUTRA) User Equipment (UE) Procedures in Idle Mode," v. 10.3.0 (2011-09), TS 36.304.
14. N. Himayat, et al., 'Heterogeneous Networking for Future Wireless Broadband Networks,' IEEE C8021610_0003r1, January 2010.
15. V. Chandrasekhar, J. Andrews, and A. Gatherer, 'Femtocell Networks: A Survey,' IEEE Communications Magazine, September 2008.
16. In-Building Small Cell Networks: Achieving High Capacity Indoors, https://www.iaria.org/conferences2021/filesICWMC21/RonyKumerSaha_Tutorial_InBuildingSmall_Summary.pdf.

Index

www.ingramcontent.com/pod-product-compliance
Lightning Source LLC
LaVergne TN
LVHW090404210926
848480LV00002B/446
9798232590192